UNIFORM RANDOM NUMBERS
Theory and Practice

THE KLUWER INTERNATIONAL SERIES IN ENGINEERING AND COMPUTER SCIENCE

DISCRETE EVENT DYNAMIC SYSTEMS

Consulting Editor

Yu-Chi Ho

Harvard University

GRADIENT ESTIMATION VIA PERTURBATION ANALYSIS, P. Glasserman
ISBN: 0-7923-9095-4

PERTURBATION ANALYSIS OF DISCRETE EVENT DYNAMIC SYSTEMS, Yu-Chi Ho and Xi-Ren Cao
ISBN: 0-7923-9174-8

PETRI NET SYNTHESIS FOR DISCRETE EVENT CONTROL OF MANUFACTURING SYSTEMS, MengChu Zhou and Frank DiCesare
ISBN: 0-7923-9289-2

MODELING AND CONTROL OF LOGICAL DISCRETE EVENT SYSTEMS, Ratnesh Kumar and Vijay K. Garg
ISBN: 0-7923-9538-7

UNIFORM RANDOM NUMBERS
Theory and Practice

by

Shu Tezuka
IBM Japan, Ltd.
1623-14, Shimo-tsuruma, Yamato-shi,
Kanagawa-ken 242, Japan

Foreword

by

Yu-Chi Ho
Harvard University

KLUWER ACADEMIC PUBLISHERS
Boston / Dordrecht / London

Distributors for North America:
Kluwer Academic Publishers
101 Philip Drive
Assinippi Park
Norwell, Massachusetts 02061 USA

Distributors for all other countries:
Kluwer Academic Publishers Group
Distribution Centre
Post Office Box 322
3300 AH Dordrecht, THE NETHERLANDS

Library of Congress Cataloging-in-Publication Data

A C.I.P. Catalogue record for this book is available
from the Library of Congress.

Printed on acid-free paper.

Printed in the United States of America

To
Hiroko and Momoko

CONTENTS

SERIES FOREWORD

In earlier forewords to the books in this series on Discrete Event Dynamic Systems (DEDS), we have dwelt on the pervasive nature of DEDS in our human-made world. From manufacturing plants to computer/communication networks, from traffic systems to command-and-control, modern civilization cannot function without the smooth operation of such systems. Yet mathematical tools for the analysis and synthesis of DEDS are nascent when compared to the well developed machinery of the continuous variable dynamic systems characterized by differential equations. The performance evaluation tool of choice for DEDS is discrete event simulation both on account of its generality and its explicit incorporation of randomness.

As it is well known to students of simulation, the heart of the random event simulation is the uniform random number generator. Not so well known to the practitioners are the philosophical and mathematical bases of generating "random" number sequence from deterministic algorithms. This editor can still recall his own painful introduction to the issues during the early 80's when he attempted to do the first perturbation analysis (PA) experiments on a personal computer which, unbeknownst to him, had a random number generator with a period of only 32,768 numbers. It is no exaggeration to say that the development of PA was derailed for some time due to this ignorance of the fundamentals of random number generation.

It is thus a pleasure to welcome Dr. Shu Tezuka's authoritative book on the subject into the series. As his preface makes clear, both theoreticians and practitioners should find the book immediately useful and indispensable.

Yu-Chi Ho
Editor

PREFACE

In the past decade, random number generation by computers has become much broader in scale and more sophisticated in the depth of its application areas, which include discrete-event simulation, VLSI pseudorandom testing, cryptography, randomized algorithms, computational physics, financial engineering, and games such as pachinko. Since almost all non-uniform random numbers are commonly produced by some transformation of uniform random variables, the computer generation of random numbers that are uniformly distributed between zero and one is regarded as the most important technology for those applications. However, at several places in the world, one can still find various instances of the use of bad uniform random number generators, and of the misuse or misunderstanding of good generators. I believe that correct and appropriate understanding of uniform random number generation will increase the efficiency and significance of its implementations and applications.

The purpose of this monograph is to provide a comprehensive exposition of major developments over the last decade in uniform random number generation for practical use. I hope that this will meet the needs of general readers who are interested in state-of-the-art algorithmic methods for generating "randomness" on a computer. Since the emphasis is on practice in relation to the theoretical background of uniform random numbers, discussion of interesting but rather philosophical questions concerning randomness itself is limited to some appropriate references. The organization of the monograph is as follows: Chapter 1 briefly overviews the current state and the growing importance of random number generation in computer science, and adds some historical remarks. In order to make the monograph as self-contained as possible, I have assembled the mathematical underpinnings of uniform random number generation under one roof in Chapter 2. The topics covered there are from three branches of number theory: modular arithmetic, geometry of numbers, and uniform distribution of sequences. Practitioners who would prefer to move on as quickly as possible to random numbers can skip this chapter or use it as a collection of theoretical backup data. The remaining chapters can be read more or less independently. Chapter 3 explains lattice structures, spectral tests, and relevant recent findings on linear congruential generators with long periods. Chapter 4 discusses the development of several random number generators, with the main focus on shift register generators, because of their practical importance. Chapter 5 summarizes various commonly used statistical tests for randomness, and discusses a certain irregularity in some types of non-uniform random variable generation due to the structure of uniform random numbers. The last chapter includes one of the most rapidly emerging topics in this area, the interesting notion of derandomization, which means making algorithms that involve randomness purely deterministic. We discuss two approaches for derandomization: low-discrepancy sequences and k-wise independent random variables. The appendix contains sample C subroutines for generating uniform random numbers and low-discrepancy sequences.

Most of the results and materials included in Chapters 3 through 6 were obtained in the course of the author's one-man project on random number generation at IBM's Tokyo Research Laboratory, from May 1989 through February 1994. I would like to express my sincere gratitude to TRL managers, especially the Director, N. Suzuki, for their encouragement and support in this five-year project. I am deeply indebted to Professor J. F. Traub for inviting me to Columbia in November 1993, where I learned that quasi-Monte Carlo methods are enormously useful for real practical problems in financial engineering. This experience strongly encouraged me to write about low-discrepancy sequences in Chapter 6. I also thank Professors R. Couture, M. Fushimi, P. L'Ecuyer, T. Tokuyama, and Y. J. Wong for their contributions to the project.

Without the generous and kind efforts of several people, this monograph would have never been published in its current form. First, I would like to express my gratitude to Professor Y. C. Ho for recommending the monograph for inclusion in the Kluwer DEDS series, and to Professors H. Niederreiter and H. Woźniakowski for their encouragement and many valuable comments, which led to considerable improvements in the manuscript. Many thanks also go to M. McDonald for his careful reading of the whole draft.

Shu Tezuka
Yamato, Kanagawa

1

INTRODUCTION

The most important questions of life are,
for the most part, really only problems of probability.
– Pierre Simon, Marquis de Laplace (1812)

Recently, the proliferation of personal computers and workstations has greatly broadened the applicability of technology for stochastic computer simulation, in which we need to model and analyze real or virtual systems in uncertain conditions. On a computer, the uncertainty in such systems is usually simulated by using random numbers. While, in the past, the purpose of simulation was to obtain qualitative understanding of the targeted system or the model being validated, the explosive growth of computing power in the last decade has made the objective much more quantitative. The so-called 'large-scale' simulations have made two kinds of estimation feasible: estimation of quantities with large variances, and highly accurate estimation of targeted variables. We now need random numbers, with high quality and long periods, that are portable and can be efficiently generated.

Usually, when we simulate a particular model containing uncertainty, we need random variables with a particular probability distribution, such as normal or exponential distribution. In practice, almost all of the algorithms for generation of these non-uniform random variables are based on certain transformations of one or more uniform random numbers. Uniform random numbers can thus be viewed as the life-blood of stochastic computer simulation. For this reason, efficient and portable generation of reliable uniform random numbers has long been an essential task for simulation modelers and analysts.

Actually, generation of uniform random numbers on a computer is a deterministic simulation of randomness by means of arithmetic operations. Therefore, sequences obtained by this kind of deterministic process are totally different from sequences of truly random numbers. Historically, the notion of "true randomness" is an idealized concept statistically purified through the modeling of some physical processes such as dice-casting or coin-flipping. While this background justifies the use of such physically generated random numbers, they are not very popular in the field of computer simulation, for two main reasons. The first is that, as stated above, those random numbers generated by using a particular physical process are also different from truly random numbers, which exist only conceptually. For example, there is no ideal dice without any bias that can simulate Bernoulli trials. The second is the inefficiency of reproducing or reusing the same sequence of such random numbers several times in the same computer or a different (remote) one. This procedure is essential for checking the simulation results obtained by other people and for debugging simulation programs. In particular, the reproducibility of random numbers is an essential element of such applications as discrete-event simulations and VLSI pseudorandom testing. In the former, it is known that variance reduction techniques such as *common random numbers* need the same stream of random numbers, to increase the efficiency with which two different systems can be compared by means of simulation. In the latter, random numbers or patterns are used as input sequences to the VLSI chips being tested, whose output sequences are compared to the reference data calculated in advance by (actual or numerical) simulation of the correct (no-fault) VLSI using the same sequence of random numbers. If we do not use the same sequence of random numbers, the idea of the testing scheme does not make sense.

In principle, it is impossible to "generate" true randomness on a computer, which is a totally deterministic machine. It is obvious that any sequences generated from deterministic processes are "pseudorandom" and not *random.* We can choose from at least two totally different (optimistic and pessimistic) ways of overcoming (or circumventing) this situation. One is a kind of engineering solution, that is, to continue to use pseudorandom numbers with some optimism; the other choice is to give up attempting to create randomness on a computer, which eventually leads us to the notion of derandomization.

For the first choice, we need to carry out a battery of statistical tests on the sequences used for our particular applications. Certainly, they should be tested in many different ways, and then improved step by step to mimic randomness ever more closely. However, we can never reach true randomness. As Marsaglia [58] says, "Every deterministic scheme for producing randomness must have applications for which it gives bad results. Only the collective experience and imagination of developers and users of random number generators will lead to a better understanding of what those applications are." Anyway, we have certainly had steady and considerable progress in this direction of research since the 1940s, the days of Neumann and Lehmer, who were the first to consider computer-generated random numbers for Monte Carlo methods. For efficient use of the resources of time and space, most algorithms have been based on a simple recurrence relation, which can be viewed as an example of a discrete dynamical system with a finite (small) memory. Two critical issues – namely, discreteness and finiteness – begin to emerge if we consider any type of algorithm for computer generation of random numbers. Discreteness implies that random numbers should be modeled as integer sequences rather than as sequences of real numbers. Finiteness makes any sequences produced on a computer periodic. For example, people who are familiar with the notion of chaos might think of chaotic (or random) behavior in some simple nonlinear discrete dynamical systems such as

$$x_{n+1} = 4x_n(1 - x_n) \text{ for } n = 1, 2, ...,$$

as a candidate for a deterministic scheme of (pseudo)random number generation. However, on a computer, we cannot simulate chaotic behavior in any sense, because of the finite precision of the numbers $x_n, n = 1, 2, ...,$. In fact, as noted in [75], the sequence often enters one of two periodic cycles of lengths 160853385 or 105364478, if double-precision (64-bit) floating-point arithmetic is used for the calculation. Hence, we should be aware that (pseudo)random number generation on a computer suffers from having finite periods. To sum up, the main problem in random number generation on a computer is how to generate quickly, with a small amount of memory, long-period sequences of integers that look like finite segments of truly random numbers.

In contrast to the above optimistic approach of adopting an engineering compromise to resolve the difficult situation by replacing truly random sequences with pseudorandom numbers are the efforts of some people to seek a much more rational, rigid, and mathematical solution. According to von Neumann [104], "It is true that a problem that we suspect of being solvable by random methods may be solvable by some rigorously defined sequence." On the basis of this view, some mathematicians (e.g., Zaremba, see [111]) have proposed an alternative approach using deterministic versions of Monte Carlo methods, which they call quasi-Monte Carlo methods. However, we should note that mathematical treatments capable of providing a complete alternative to the whole theory of Monte Carlo methods are almost infeasible, and that the area that can be covered by quasi-Monte Carlo methods on a sound mathematical basis is quite limited at the time of writing, although there have been considerable and continuous efforts to broaden the range of applications.

The organization of this monograph is as follows: Chapter 2 summarizes the mathematical preliminaries from three branches of number theory: modular arithmetic, geometry of numbers, and uniform distribution of sequences. Since the main objective of this chapter is to assemble the mathematical underpinnings of uniform random number generation under one roof, the reader who would like to learn as quickly as possible about particular algorithms for random numbers can skip this chapter or use it as a supplementary source of theoretical background material for random number generation. The first section overviews modular arithmetic of integers or polynomials, including the Chinese remainder theorem, finite fields, and relevant topics such as formal Laurent series and linear feedback shift register sequences. Section 2.2 describes fundamental results from the geometry of numbers for integer lattices as well as polynomial lattices, and also gives lattice basis reduction algorithms, such as the Lenstra-Lenstra-Lovász (LLL) algorithm for integer lattices and the Lenstra algorithm for polynomial lattices. In Section 2.3, we first overview uniform distribution of infinite sequences of integers or polynomials, and summarize important topics concerning irregularities of distribution, such as the definitions of discrepancy, low-discrepancy sequences, and Halton sequences.

The remaining chapters can be read more or less independently. In each chapter, I have added proofs for almost all of my original results. Where I have not done so, I have given the idea of the proof or provided an appropriate reference. Chapter 3 contains recent results on the efficient generation of linear congruential sequences with long periods. The lattice structure of linear congruential sequences is briefly overviewed, and the spectral test developed for examining such lattices is explained in detail. The recently proposed Marsaglia-Zaman and Wichmann-Hill schemes are fully described and are shown to allow fast generation of linear congruential sequences with long periods.

Chapter 4, entitled "Beyond Linear Congruential Generators," is divided into three parts. In the first, Tausworthe and Lewis-Payne random numbers are formulated as linear congruential sequences with respect to polynomial arithmetic over finite fields, in which the transformation from finite fields to real numbers in $[0,1)$ behaves as a nonlinear operation. This class of random number generators can be implemented by using linear feedback shift register sequences, and is thus of great practical importance to the generation of uniform random numbers. We describe in detail a theoretical test for analyzing the uniform distribution of this class of sequences, and apply the test to many currently-used generators with various period lengths. In the second part, we discuss inversive congruential generators. This class of generators has been defined as a modification of linear congruential generators obtained by using multiplicative inversion. The nonlinearity due to the inversion destroys the integer lattice structure of the conventional linear congruential sequences. The last part concerns the random binary sequences that are being investigated in the field of cryptography, where the unpredictability of sequences is a key issue. The questions of how to define and how to generate such unpredictable sequences are discussed, and two types of proposal are summarized. One approach is based on cryptographical secureness using computational complexity theory, and the other on the concept of linear complexity, which is closely related to Kolmogorov-Chaitin's definition of randomness.

Chapter 5 summarizes various statistical tests for randomness. In the first half, test procedures, general and specific, are described briefly. We then suggest that several schemes of non-uniform random variable generation can be used for testing the hypothesis of the uniformity of sequences in $[0, 1)$, and illustrate this idea by using the so-called Neave effect for the Box-Muller method of generating normal random variables, the most frequently used non-uniform random variables.

The last chapter covers two important topics closely related to derandomization. First, we discuss very recent developments in low-discrepancy sequences, mainly the generalized Niederreiter sequences, which are a key technology for quasi-Monte Carlo methods. The practical issue of implementation for fast generation of low-discrepancy sequences is also discussed. Second, as an important related topic in discrepancy theory as well as in uniform random numbers, we give an overview of recent results concerning k-wise independent random numbers. Their formal definitions are given along with concrete construction methods, and are followed by the notion of almost k-wise independence. We also describe how to generate almost k-wise independent random variables by using linear feedback shift register sequences.

In the appendix, we provide three sample programs written in the ANSI C programming language: two subroutines for uniform random numbers, CombLS2 and CombMRG, which were implemented on the basis of the results described in Section 4.1, for uniform random numbers, and another for low-discrepancy sequences, GFaure, which was implemented according to the results obtained in Section 6.1.

For precision, some people call those numbers deterministically generated on a computer "pseudorandom numbers," reserving the word "random" for the output of some really *random* process, but following Knuth's usage, we simply call them "random numbers" throughout this monograph except when we need to be more precise.

* * *

Notes and Remarks

A philosophical discussion and mathematical definition of randomness can be found in Knuth [40] and in Chaitin [18]. Zaremba has discussed the mathematical basis of Monte Carlo and quasi-Monte Carlo methods [111]. An encyclopedic summary of computer algorithms for many kinds of non-uniform random numbers, almost all of which are based on transformation using uniform random variables, is given by Devroye [24].

Applications of random numbers to discrete-event simulations are found in publications by Bratley et al. [16], Ho and Cao [35], and many others. VLSI pseudorandom testing has become a major application of random numbers, both in software and hardware [103]. Random number generation is also among the most important topics in cryptography. Partial lists of references specific to such applications can be found in Section 4.3. Applications of random numbers to the design of algorithms are discussed in Motwani and Raghavan [63]. Kalos and Whitlock [39] discuss topics related to the use of random numbers in the field of computational physics, including many applications such as Ising model simulation.

2

PRELIMINARIES FROM NUMBER THEORY

Anyone who considers arithmetical
methods of producing random digits
is, of course, in a state of sin.
– John von Neumann (1951)

This chapter summarizes mathematical preliminaries from three branches of number theory: modular arithmetic, geometry of numbers, and uniform distribution of sequences, all of which constitute a theoretical foundation for the generation of uniform random numbers. Since the main objective of this chapter is to make the monograph self-contained, readers who would like to learn as quickly as possible about particular algorithms for random numbers can skip this chapter and consult as necessary those elements of theory that are needed for the problems they are studying.

2.1 MODULAR ARITHMETIC

Automatic digital computers can represent internally only a finite set of numbers, and thus they cannot deal with the set of all integers, because this is infinite and most of its elements cannot be represented on a computer. However, we can perform certain arithmetic operations exactly if the operands are integers bounded by the word-size of a computer. If we use integer arithmetic and reduce the results *modulo* m, we call it (integer) modular arithmetic (or residue arithmetic), where the integer $m > 1$ is called the modulus of the arithmetic.

2.1.1 Integer Arithmetic

In this section, we shall briefly overview several results from elementary number theory which are relevant to the developments in the succeeding chapters. For more details and proofs, see, for example, Hardy and Wright [33].

DEFINITION **2.1** Let a, b, and m be integers with $m > 1$. If $a - b$ is divisible by m, we write

$$a \equiv b \pmod{m},$$

which reads *a is congruent to b modulo m*. □

The relation $a \equiv b \pmod{m}$ is called *congruence*. The following are several properties of congruence:

Property **2.1** *Let a, b, c, d, x, y, and m be integers with $m > 1$.*

- *If $a \equiv b \pmod{m}$ and $b \equiv c \pmod{m}$, then $a \equiv c \pmod{m}$.*
- *If $a \equiv b \pmod{m}$ and $c \equiv d \pmod{m}$, then $ac \equiv bd \pmod{m}$.*
- *If $a \equiv b \pmod{m}$ and $c \equiv d \pmod{m}$, then $ax + cy \equiv bx + dy \pmod{m}$.*

Let $I_m = \{0, 1, ..., m-1\}$. By using the congruence relation, we can map each integer x onto an integer $r \in I_m$. We call r the (least nonnegative) residue of x modulo m, and denote it by the notation $|x|_m$. The following properties are useful:

Property **2.2** *Let a, b, and m be integers with $m > 1$.*

- *$|a|_m$ is unique.*
- *$|a|_m = |b|_m$ if and only if $a \equiv b \pmod{m}$.*
- *$|km|_m = 0$ for every integer k.*
- $|a + b|_m = ||a|_m + |b|_m|_m = ||a|_m + b|_m = |a + |b|_m|_m$.
- $|ab|_m = ||a|_m|b|_m|_m = ||a|_m b|_m = |a|b|_m|_m$.

Theorem 2.1 *The system $(I_m, +, *)$, where + and * denote addition and multiplication modulo m, respectively, constitutes a finite commutative ring with identity.*

The following is important:

Theorem 2.2 *The finite commutative ring $(I_m, +, *)$ is isomorphic to a finite field with m elements, $GF(m)$, if and only if m is a prime.*

If m is not prime, a non-zero element of $(I_m, +, *)$ may or may not have a multiplicative inverse. The next theorem determines whether one exists.

Theorem 2.3 *Let a be an integer. Then there exists a unique integer $b \in I_m$ that satisfies $|ab|_m = |ba|_m = 1$, if and only if $|a|_m \neq 0$ and $\gcd(a, m) = 1$.*

In order to deal with a very large integer on a computer, we use more than one modulus. The idea is that modular arithmetic using several moduli is equivalent to modular arithmetic using the least common multiple of the moduli as a single modulus. The following famous theorem is known as the Chinese remainder theorem:

Theorem 2.4 *Let $m_1, ..., m_J$ be distinct moduli, and let $a_1, ..., a_J$ be integers. Assume that the moduli are pairwise coprime, namely, that $\gcd(m_i, m_j) = 1$ for $1 \leq i \neq j \leq J$. Then $x \equiv a_j \pmod{m_j}, j = 1, ..., J$, if and only if $x \equiv a_1 M_{-1} M_1 + \cdots + a_J M_{-J} M_J \pmod{M}$, where $M = \prod_{j=1}^{J} m_j$, $M_j = M/m_j$, and M_{-j} is the multiplicative inverse of M_j modulo m_j, i.e., $M_{-j} M_j \equiv 1 \pmod{m_j}$.*

The following property is useful in arithmetic using several moduli:

Property **2.3** *Let $x \equiv a_j \pmod{m_j}$, $j = 1, ..., J$, and $y \equiv b_j \pmod{m_j}$, $j = 1, ..., J$. Then $x \pm y \equiv a_j \pm b_j \pmod{m_j}, j = 1, ..., J$, and $xy \equiv a_j b_j \pmod{m_j}$, $j = 1, ..., J$.*

From now on, we summarize several definitions and results concerning the (multiplicative) order of an element of I_m, which are useful in analyzing the period length of random number sequences.

DEFINITION **2.2** Let a and m be coprime, that is, let $\gcd(a,m) = 1$. The (multiplicative) order of a modulo m is defined as the smallest positive integer e such that $a^e = 1 \pmod{m}$. □

The next classic theorem is called Fermat's theorem:

Theorem 2.5 *Let m be a prime and let a be an integer not divisible by m. Then we have*

$$a^{m-1} \equiv 1 \pmod{m}.$$

Note that the theorem says that $m-1$ is divisible by the order of a.

The Euler function is very popular and useful in number theory:

DEFINITION **2.3** For an integer $m > 0$, the Euler function, $\phi(m)$, is defined as the number of positive integers r smaller than m that are coprime to m, where $\phi(1)$ is defined to be 1. □

Euler's famous theorem is more general than Fermat's:

Theorem 2.6 *Let a and m be coprime. Then we have*

$$a^{\phi(m)} \equiv 1 \pmod{m}.$$

Next, we are concerned with what is the largest order of an integer modulo a prime.

DEFINITION **2.4** Let m be a prime. The number a is called a primitive root modulo m if the order of a is equal to $m-1$. □

The following theorem asserts the existence of a primitive root:

Theorem 2.7 *Let m be a prime. Then there exist $\phi(m-1)$ distinct primitive roots modulo m.*

DEFINITION **2.5** Let m have the primitive root g, and let k be coprime to m. We define the index of k modulo m as the smallest positive integer t for which

$$g^t \equiv k \pmod{m},$$

and write $t = \mathrm{ind}_g^{(m)}(k)$. □

Note that $\mathrm{ind}_g^{(m)}(a) = \mathrm{ind}_g^{(m)}(b)$ if and only if $a \equiv b \pmod{m}$. We then have the following [51, Theorem 1.15]:

Theorem 2.8 *Let m be a prime, let e be the order of a modulo m, and let g be a primitive root modulo m. Then we have*

$$e = \frac{m-1}{d},$$

where $d = \gcd(\mathrm{ind}_g^{(m)}(a), m-1)$, and the number of integers in I_m whose order is equal to e is $\phi(e)$.

What if m is not a prime, but a composite number? Since the system $(I_m, +, *)$ is not a field, there are elements that have no multiplicative inverses. Here, we should note that every element a that satisfies $a^e = 1 \pmod{m}$ for some nonzero e has a multiplicative inverse a^{e-1}. Thus, only values of a that satisfy $\gcd(a, m) = 1$ have nonzero orders. The following is due to Knuth [40]:

DEFINITION **2.6** Let m be a positive integer larger than 1. Then a is called a primitive element modulo m if the order of a is maximal. □

Theorem 2.9 *Let $\lambda(m)$ denote the maximal order modulo m. Then $\lambda(2) = 1$, $\lambda(4) = 2$, $\lambda(2^e) = 2^{e-2}$ for $e \geq 3$, and $\lambda(p^e) = p^{e-1}(p-1)$ for a prime $p > 2$. For $M = p_1^{e_1} \cdots p_J^{e_J}$, we have $\lambda(M) = \mathrm{LCM}(\lambda(p_1^{e_1}), ..., \lambda(p_J^{e_J}))$, where $p_1, ..., p_J$ are distinct primes.*

Next, we give the definition of continued fractions, which play an important role in the following chapters, and also have an important connection with modular arithmetic, for example, in the calculation of $\gcd(a, m)$ as well as the multiplicative inverse of a modulo m. A simple continued fraction expansion of a rational number has the form

$$a_0 + \cfrac{1}{a_1 + \cfrac{1}{a_2 + \ddots}},$$

which is customarily written as $[a_0; a_1, a_2, ...]$, where $a_i, i = 0, 1, ...$, are positive integers. A finite simple continued fraction is $[a_0; a_1, a_2, ..., a_m]$. If we break off a simple continued fraction after $j \leq m$, then $P_j/Q_j = [a_0; a_1, a_2, ..., a_j]$ is called the approximating fraction or convergent of order j. Here P_j and Q_j are coprime integers, and satisfy the recurrence relation, for $j = 1, 2, ..., m$,

$$P_j = a_j P_{j-1} + P_{j-2},$$

with $P_0 = a_0$ and $P_{-1} = 1$. Further,

$$Q_j = a_j Q_{j-1} + Q_{j-2},$$

with $Q_0 = 1$ and $Q_{-1} = 0$. If we add the condition $a_j \neq 1$, then P_j/Q_j has a unique representation $[a_0; a_1, a_2, ..., a_j]$. For example, $1/2$ has the representation $[0; 2]$, not $[0; 1, 1]$. Here we present an important property:

$$P_j Q_{j-1} - P_{j-1} Q_j = (-1)^{j+1}.$$

This implies that as the index j increases, the approximation approaches the true value of the fraction from above and below alternately.

The following examples show the continued fraction expansions of some familiar real numbers:

$$\pi = [3; 7, 15, 1, 292, 1, 1, 1, 2, 1, 3, ...],$$

$$e = [2; 1, 2, 1, 1, 4, 1, 1, 6, 1, 1, 8, ...],$$

$$\sqrt{2} = [1; \bar{2}],$$

where $\bar{a_i}$ means the infinite repetition of a_i, i.e., a periodic continued fraction. Another interesting periodic continued fraction is the famous golden ratio, i.e., $\frac{1+\sqrt{5}}{2} = [1; \bar{1}]$. The Fibonacci numbers, defined by the recurrence relation

$$F_i = F_{i-1} + F_{i-2}, \text{ with } F_0 = 0 \text{ and } F_1 = 1,$$

has a close connection with the golden ratio, since for all $i \geq 1$, we have

$$\frac{F_{i+1}}{F_i} = [1; \overbrace{1, ..., 1}^{i-1}].$$

Finally, we summarize fundamental results about arithmetical functions, which are among the most important tools in number theory, and which include the divisor function d, the Möbius function μ, and the already defined Euler function ϕ. The divisor function $d(n)$ is defined as the number of distinct divisors of n, that is, for $n = \prod_{i=1}^{k} p_i^{e_i}$,

$$d(n) = \prod_{i=1}^{k}(e_i + 1),$$

where $p_1, ..., p_k$ are distinct primes. Note that $d(1) = 1$.

Property **2.4** *We have*

- $d(mn) = d(m)d(n)$ *for* m *and* n *with* $\gcd(m, n) = 1$.
- $d(n) = \sum_{d|n} 1$.

The Möbius function is defined as follows:

Definition **2.7** μ is called the Möbius function if it satisfies

(i) $\mu(1) = 1$;

(ii) $\mu(n) = 0$, if n has a squared factor;

(iii) $\mu(n) = (-1)^k$, if n is the product of k distinct primes. □

Property **2.5** *We have*

- $\mu(mn) = \mu(m)\mu(n)$ *for* m *and* n *with* $\gcd(m, n) = 1$.
- $\mu(mn) = 0$, *if* $\gcd(m, n) > 1$.
- $\sum_{d|n} \mu(d) = 0$, *if* $n > 1$.

The following theorem is called the Möbius transform:

Theorem 2.10 *Let f be a function on positive integers. Consider*

$$F(n) = \sum_{d|n} f(d).$$

Then we have

$$f(n) = \sum_{d|n} \mu\left(\frac{n}{d}\right) F(d) = \sum_{d|n} \mu(d) F\left(\frac{n}{d}\right).$$

Here, F is called the Möbius transform of f, and f is called the inverse Möbius transform of F. We have a noteworthy relation between the divisor function and the Euler function.

Theorem 2.11

$$\sum_{i=1}^{d(n)} \phi\left(\frac{n}{d_i}\right) = n,$$

where d_i is a divisor of n. This relation can be written as follows:

$$\sum_{d|n} \phi\left(\frac{n}{d}\right) = \sum_{d|n} \phi(d) = n.$$

In other words, $F(n) = n$ is the Möbius function of the Euler function. Thus, we have

$$\phi(n) = \sum_{d|n} \mu\left(\frac{n}{d}\right) d = n \sum_{d|n} \frac{\mu(d)}{d}.$$

This is a relation between the Euler and Möbius functions.

2.1.2 Polynomial Arithmetic

Polynomials over finite fields, particularly over $GF(2)$, play an important role in digital signal processing, cryptology, and random number generation. Let q be a prime power. We denote a ring of polynomials over the finite field $GF(q)$ by $GF[q, z]$.

DEFINITION **2.8** Let $f(z) \in GF[q, z]$ be of positive degree r, and let $GF(q')$ be a field which contains $GF(q)$ as a subfield. Then $f(z)$ is said to split in $GF(q')$

if $f(z)$ can be written as a product of linear factors in $GF[q', z]$, that is, if there exist elements $\alpha_1, ..., \alpha_r \in GF(q')$ such that

$$f(z) = a(z - \alpha_1) \cdots (z - \alpha_r),$$

where a is the leading coefficient of $f(z)$. The field $GF(q')$ is called a splitting field of $f(z)$ over $GF(q)$ if $f(z)$ splits in $GF(q')$ and if, moreover, no proper subfield of $GF(q')$ contains $\alpha_1, ..., \alpha_r$. □

We extend Definition 2.2 to general finite fields.

DEFINITION **2.9** Let α be a nonzero element of $GF(q)$. The (multiplicative) order of α in $GF(q)$ is defined as the smallest positive integer e such that $\alpha^e = 1$. □

We consider factorization of $z^N - 1$ over a finite field $K = GF(q)$. The roots of $z^N - 1$ are the Nth roots of unity in the splitting field, denoted by $K^{(N)}$, of $z^N - 1$. We call a root of $z^N - 1$ a primitive (Nth) root of unity over K if the order of the root is equal to N. Let W_N be a primitive root of $z^N - 1$. Then all of $W_N^n, n = 0, 1, ..., N - 1$, are distinct. Thus, the factorization becomes

$$z^N - 1 = \prod_{n=0}^{N-1} (z - W_N^n).$$

Since this factorization is unique, all the primitive roots are given by W_N^n, for n with $\gcd(n, N) = 1$. By taking the products of all of them, we define cyclotomic polynomials as follows:

DEFINITION **2.10** Let W_n be a primitive nth root of unity over $GF(q)$, and assume that n is not divisible by the characteristic[1] of $GF(q)$. Then the polynomial

$$C_n(z) = \prod_{\substack{1 \le d \le n \\ \gcd(d,n)=1}} (z - W_n^d),$$

is called the nth cyclotomic polynomial over $GF(q)$. □

The important properties of cyclotomic polynomials are given below.

[1]The characteristic of $GF(q)$ is defined as the smallest positive integer p such that $p\alpha = 0$ for every $\alpha \in GF(q)$.

Theorem 2.12 *Let p be the characteristic of $GF(q)$, and let n be a positive integer not divisible by p. Then*

$$z^n - 1 = \prod_{d|n} C_d(z);$$

$$C_n(z) = \prod_{d|n} (z^d - 1)^{\mu(n/d)} = \prod_{d|n} (z^{n/d} - 1)^{\mu(d)}; \tag{2.1}$$

and the coefficients of $C_n(z)$ belong to $GF(p)$.

The formula (2.1) can be derived based on the inverse Möbius transform [51, Theorem 3.27]. Hence, the degree of $C_n(z)$ is

$$\deg(C_n(z)) = \sum_{d|n} \mu\left(\frac{n}{d}\right) d = \phi(n).$$

Next, we introduce the definitions of irreducible and primitive polynomials:

DEFINITION **2.11** A polynomial $f(z) \in GF[q, z]$ is said to be irreducible over $GF(q)$ (or irreducible in $GF[q, z]$) if $f(z)$ has positive degree and $f(z) = b(z)c(z)$ with $b(z), c(z) \in GF[q, z]$ implies that either $b(z)$ or $c(z)$ is a constant. □

Note that 'irreducible' polynomials in $GF[q, z]$ correspond to 'prime' numbers in the integer ring, and that both have many common algebraic features. We have important theorems.

Theorem 2.13 *If $f(z)$ is an irreducible polynomial in $GF[q, z]$ of degree r, then $f(z)$ has a root $\alpha \in GF(q^r)$. Furthermore, all the roots of $f(z)$ are simple and are given by the r distinct elements $\alpha, \alpha^q, \alpha^{q^2}, ..., \alpha^{q^{r-1}}$ of $GF(q^r)$.*

Corollary 2.1 *Let $f(z)$ be an irreducible polynomial in $GF[q, z]$ of degree r. Then the splitting field of $f(z)$ over $GF(q)$ is given by $GF(q^r)$.*

Theorem 2.14 *For every finite field $GF(q)$ and every positive integer r, the product of all monic irreducible polynomials in $GF[q, z]$ whose degrees divide r is equal to $z^{q^r} - z$.*

Corollary 2.2 *Let $N_q(d)$ be the number of monic irreducible polynomials in $GF[q,z]$ of degree d. We then have, for all positive integers r,*

$$q^r = \sum_{d|r} dN_q(d).$$

By using the inverse Möbius transform, we have

Corollary 2.3 *The number $N_q(r)$ of monic irreducible polynomials in $GF[q,z]$ of degree r is given by*

$$N_q(r) = \frac{1}{r}\sum_{d|r} \mu\left(\frac{r}{d}\right) q^d = \frac{1}{r}\sum_{d|r} \mu(d) q^{r/d}.$$

Let $I(q,r;z)$ be the product of all monic irreducible polynomials in $GF[q,z]$ of degree r. Then the above results give

$$I(q,r;z) = \prod_{d|r} (z^{q^d} - z)^{\mu(r/d)} = \prod_{d|r} (z^{q^{r/d}} - z)^{\mu(d)}.$$

The following theorem is important:

Theorem 2.15 *For $r > 1$, we have*

$$I(q,r;z) = \prod_{d} C_d(z),$$

where the product is extended over all positive divisors d of $q^r - 1$ for which r is the multiplicative order of q modulo d.

Next, we define the order of polynomials.

DEFINITION **2.12** Let $f(z) \in GF[q,z]$ be a nonzero polynomial. If $f(0) \neq 0$, then the least positive integer e for which $f(z)$ divides $z^e - 1$ is called the order of $f(z)$ and is denoted by $\mathrm{ord}(f)$. If $f(0) = 0$, then $f(z) = z^h g(z)$, where h is a positive integer and $g(0) \neq 0$ with $g(z) \in GF[q,z]$; $\mathrm{ord}(f)$ is then defined to be $\mathrm{ord}(g)$. □

Note that the order of nonzero constant polynomials is equal to 1, since these polynomials divide $z - 1$. Also, $f(z) = z$ over $GF(2)$ has the order 1.

Theorem 2.16 *The order of an irreducible polynomial over $GF(q)$ of degree r divides $q^r - 1$.*

For the case of non-irreducible polynomials, we have

Theorem 2.17 *Let $m_1(z), ..., m_J(z)$ be pairwise coprime nonzero polynomials in $GF[q, z]$, and denote $M(z) = m_1(z) \cdots m_J(z)$. Then* $\text{ord}(M) = \text{LCM}(\text{ord}(m_1), ..., \text{ord}(m_J))$.

We now give the definition of primitive polynomials:

DEFINITION **2.13** A monic polynomial $f(z)$ with $f(0) \neq 0$ in $GF[q, z]$ of degree $r \geq 1$ is called a primitive polynomial over $GF(q)$ if its order is $q^r - 1$. □

The condition $f(0) \neq 0$ is necessary in order to exclude the non-primitive polynomial $f(z) = z$ over $GF(2)$. The definition leads to the following:

Theorem 2.18 *The number of primitive polynomials in $GF[q, z]$ of degree r is given by $\phi(q^r - 1)/r$.*

Modular arithmetic with respect to polynomials over finite fields is similar to that of integers. For two polynomials $f(z)$ and $M(z)$ in $GF[q, z]$, we have a unique factorization

$$f(z) = q(z)M(z) + g(z),$$

which is represented by

$$f(z) \equiv g(z) \pmod{M(z)},$$

where $\deg(g) < \deg(M)$. The following theorems are fundamental:

Theorem 2.19 *For a polynomial $M(z)$ in $GF[q, z]$ of degree r, the residue class ring $GF[q, z]/(M)$ is a field isomorphic to $GF(q^r)$ if and only if $M(z)$ is irreducible, where (M) denotes the ideal of $GF[q, z]$ generated by $M(z)$.*

Theorem 2.20 *Let α be a root of an irreducible polynomial $M(z)$ in $GF[q, z]$ of degree r. Then $\{1, \alpha, ..., \alpha^{r-1}\}$ is a basis of $GF(q^r)$ over $GF(q)$.*

Therefore, if we let α be a root of an irreducible polynomial $M(z) \in GF[q, z]$, then every element of $GF(q^r)$ can be represented as a polynomial $f(\alpha)$, where $\deg(f) < \deg(M)$. We now have

Theorem 2.21 *Let $M(z)$ be an irreducible polynomial of degree r and with $M(0) \neq 0$. Let α be a root of $M(z)$. Then $\mathrm{ord}(M)$ is equal to the (multiplicative) order of α in $GF(q^r)$.*

There is also the Chinese remainder theorem for polynomials. For given polynomials $y_1(z), ..., y_J(z)$, we have a unique polynomial $y(z)$ satisfying

$$y(z) \equiv y_j(z) \pmod{m_j(z)}, \ j = 1, 2, ..., J$$

such that $0 \leq \deg(y(z)) < \sum_{j=1}^{J} \deg(m_j(z))$, provided that the monic polynomials $m_1(z), ..., m_J(z)$ are pairwise coprime. An explicit form of $y(z)$ is given as follows: Denote $M(z) = \prod_{j=1}^{J} m_j(z)$, and $M_j(z) = M(z)/m_j(z)$. Let $M_{-j}(z)$ be the multiplicative inverse of $M_j(z)$ modulo $m_j(z)$. Then

$$y(z) \equiv \sum_{j=1}^{J} M_{-j}(z) M_j(z) y_j(z) \pmod{M(z)}.$$

The polynomial version of the Euler function is necessary in Chapter 4.

Definition **2.14** For a polynomial $M(z) \in GF[q, z]$, the polynomial version of the Euler function, $\Phi_q(M)$, is defined as the number of polynomials in $GF[q, z]$ that are of smaller degree than $M(z)$ and are coprime to $M(z)$. Define $\Phi_q(M) = 1$ if $\deg(M) = 0$. □

The following result is fundamental [51, Lemma 3.69]:

Theorem 2.22 *Let $m_j(z), j = 1, ..., J$ be distinct monic irreducible polynomials in $GF[q, z]$, and let $e_j, j = 1, ..., J$, be positive integers. Denote $r_j = \deg(m_j)$ and $a \neq 0 \in GF(q)$. If $M(z) = a \prod_{j=1}^{J} m_j(z)^{e_j}$ and $\deg(M) \geq 1$, then*

$$\Phi_q(M) = q^r (1 - q^{-r_1}) \cdots (1 - q^{-r_J}),$$

where $r = \sum_{j=1}^{J} e_j r_j$.

An analogue of Euler's famous theorem is given as follows:

Theorem 2.23 *Let $g(z)$ and $M(z)$ be polynomials in $GF[q,z]$ with $\gcd(g,M) = 1$. Then we have*

$$g(z)^{\Phi_q(M)} \equiv 1 \pmod{M(z)}.$$

Following Niederreiter [71], we let q be a prime power and let $GF\{q,z\}$ be the field of formal Laurent series

$$S(z) = \sum_{j=w}^{\infty} x_j z^{-j}$$

in z^{-1}, where all $x_j \in GF(q)$ and w is an arbitrary integer with $x_w \neq 0$. Define the discrete exponential valuation ν on $GF\{q,x\}$ as follows: For $S(z) \neq 0$ we put $\nu(S) = -w$ if w is the least index with $x_w \neq 0$, and for $S(z) = 0$ we put $\nu(S) = -\infty$. Note that the field $GF\{q,z\}$ contains the field of rational functions over $GF(q)$ as a subfield. We have the following properties for $S_1(z), S_2(z) \in GF\{q,z\}$:

$$\nu(S_1 S_2) = \nu(S_1) + \nu(S_2),$$

$$\nu(S_1 + S_2) \le \max(\nu(S_1), \nu(S_2)),$$

$$\nu(S_1 + S_2) = \max(\nu(S_1), \nu(S_2)) \text{ if } \nu(S_1) \neq \nu(S_2).$$

For $P(z), Q(z) \in GF[q,z]$ with $Q(z) \neq 0$, we have $\nu(P/Q) = \deg(P) - \deg(Q)$.

Every $S(z) \in GF\{q,z\}$ has a unique continued fraction expansion of the form

$$S(z) = A_0(z) + \cfrac{1}{A_1(z) + \cfrac{1}{A_2(z) + \ddots}} =: [A_0(z); A_1(z), A_2(z), \ldots],$$

where $A_j(z) \in GF[q,z]$ and $\deg(A_j(z)) \ge 1$ for $j \ge 1$. This expansion is finite for a rational $S(z)$, and infinite for irrational $S(z)$. The polynomials $A_j(z), j \ge 0$, are obtained recursively by the following procedure: Let $[S(z)]$ denote the polynomial part of $S(z)$.

$$A_0(z) = [S(z)],\ B_0(z) = S(z) - [S(z)],$$

$$A_{j+1}(z) = \left[\frac{1}{B_j(z)}\right],\ B_{j+1}(z) = \frac{1}{B_j(z)} - A_{j+1}(z) \text{ for } j \ge 0,$$

which can be continued as long as $B_j(z) \neq 0$. If the continued fraction expansion is broken off after the term $A_j(z)$, we get the rational convergent

$P_j(z)/Q_j(z)$. The polynomials $P_j(z)$ and $Q_j(z)$ can be calculated recursively by

$$P_{-1}(z) = 1, P_0(z) = A_0(z), P_j(z) = A_j(z)P_{j-1}(z) + P_{j-2}(z) \text{ for } j \geq 1,$$

$$Q_{-1}(z) = 0, Q_0(z) = 1, Q_j(z) = A_j(z)Q_{j-1}(z) + Q_{j-2}(z) \text{ for } j \geq 1.$$

We then have

$$\deg(Q_j(z)) = \sum_{m=1}^{j} \deg(A_m(z)) \text{ for } j \geq 1,$$

where for a rational $S(z)$ we interpret $\deg(A_j(z)) = \deg(Q_j(z)) = \infty$ whenever $A_j(z)$ and $Q_j(z)$ do not exist. The following results are important:

Theorem 2.24 *For $j \geq 0$, we have*

$$P_j(z)Q_{j-1}(z) - Q_j(z)P_{j-1}(z) = (-1)^{j+1},$$

where $P_j(z)/Q_j(z)$ is the jth approximate fraction of $S(z) \in GF\{q, z\}$.

Theorem 2.25 *For $j \geq 0$, we have*

$$\nu\left(S(z) - \frac{P_j(z)}{Q_j(z)}\right) = -\deg(Q_j(z)) - \deg(Q_{j+1}(z)),$$

where $P_j(z)/Q_j(z)$ is the jth approximate fraction of $S(z) \in GF\{q, z\}$.

Theorem 2.26 *For any $n \geq 1$ and $S(z) \in GF\{q, z\}$, let $S_n(z)$ be any element in $GF\{q, z\}$ such that $\nu(S_n(z) - S(z)) = -n - 1$. Then there exists $j \geq 0$ uniquely determined by the condition*

$$\deg(Q_{j-1}(z)) + \deg(Q_j(z)) \leq n < \deg(Q_j(z)) + \deg(Q_{j+1}(z)).$$

Finally, we summarize the materials concerning linear recurring sequences. We start with the definition of *trace*, $\mathrm{Tr}_{F/K}(\alpha)$.

DEFINITION **2.15** For $\alpha \in F = GF(q^r)$ and $K = GF(q)$, the trace $\mathrm{Tr}_{F/K}(\alpha)$ of α over K is defined by

$$\mathrm{Tr}_{F/K}(\alpha) = \alpha + \alpha^q + \cdots + \alpha^{q^{r-1}}.$$

If q is a prime, then $\mathrm{Tr}_{F/K}(\alpha)$ is called the absolute trace of α and is simply denoted by $\mathrm{Tr}_F(\alpha)$. □

From Theorem 2.13, the roots of an irreducible polynomial $M(z)$ of degree r over $GF(q)$ are given by $\alpha, \alpha^q, ..., \alpha^{q^{r-1}}$. Hence

$$\begin{aligned} M(z) &= z^r - a_{r-1}z^{r-1} - \cdots - a_0 & (2.2)\\ &= (z-\alpha)\cdots(z-\alpha^{q^{r-1}}), & (2.3)\end{aligned}$$

and a comparison of coefficients shows that

$$\mathrm{Tr}_{F/K}(\alpha) = a_{r-1}.$$

The following properties are very useful:

Property **2.6** *Let $F = GF(q^r)$ and $K = GF(q)$. Then the trace $\mathrm{Tr}_{F/K}$ satisfies the following properties:*

- $\mathrm{Tr}_{F/K}(\alpha+\beta) = \mathrm{Tr}_{F/K}(\alpha) + \mathrm{Tr}_{F/K}(\beta)$ *for all* $\alpha, \beta \in F$;
- $\mathrm{Tr}_{F/K}(c\alpha) = c\mathrm{Tr}_{F/K}(\alpha)$ *for all* $c \in K, \alpha \in F$;
- $\mathrm{Tr}_{F/K}$ *is a linear transformation from F onto K, where both F and K are viewed as vector space over K;*
- $\mathrm{Tr}_{F/K}(a) = ra$ *for all* $a \in K$;
- $\mathrm{Tr}_{F/K}(\alpha^q) = \mathrm{Tr}_{F/K}(\alpha)$ *for all* $\alpha \in F$.

Let r be a positive integer. A sequence $x_1, x_2, ...$ of elements of $GF(q)$ is called an rth-order linear recurring sequence if it satisfies the recurrence relation

$$x_{n+r} = a_{r-1}x_{n+r-1} + \cdots + a_0 x_n \text{ for } n = 1, 2, ...,$$

with coefficients $a_0, ..., a_{r-1} \in GF(q)$. Here, $x_1, ..., x_r$ are initial values for the recursion, and they uniquely determine the sequence $x_n, n = 1, 2, ...$, which is purely periodic if $a_0 \neq 0$. The polynomial $M(z) = z^r - a_{r-1}z^{r-1} - \cdots - a_0$ is called the characteristic polynomial of the recurrence relation as well as of the linear recurring sequence. If the characteristic polynomial is primitive, the linear recurring sequence has a maximal period of $q^r - 1$ provided that the initial values $x_1, ..., x_r$ are not all zeros. We have an important result:

Theorem 2.27 *Let $x_n, n = 1, 2, ...$, be an rth-order linear recurring sequence in $K = GF(q)$ whose characteristic polynomial $M(z)$ is irreducible over K. Let*

α be a root of $M(z)$ in $F = GF(q^r)$. Then there exists a uniquely determined $\gamma \in F$ such that

$$x_n = \mathrm{Tr}_{F/K}(\gamma \alpha^n)$$

for $n = 1, 2, \ldots.$

Below, we illustrate the matrix representation of linear recurring sequences, which will be very useful in the next chapters. Let C be the $r \times r$ companion matrix of the characteristic polynomial $M(z) = z^r - a_{r-1}z^{r-1} - \cdots - a_1 z - a_0$, namely,

$$C = \begin{pmatrix} 0 & 0 & 0 & \cdots & a_0 \\ 1 & 0 & 0 & \cdots & a_1 \\ & \cdots & \cdots & \cdots & \\ & \cdots & \cdots & \cdots & \\ 0 & \cdots & 1 & 0 & a_{r-2} \\ 0 & \cdots & 0 & 1 & a_{r-1} \end{pmatrix},$$

and let $\mathbf{x}_1 = (x_1, \ldots, x_r)$ be a non-zero binary vector. Then $\mathbf{x}_n = (x_n, \ldots, x_{n+r-1})$, $n = 1, 2, \ldots$, are written as follows:

$$\mathbf{x}_1, \mathbf{x}_1 C, \ldots, \mathbf{x}_1 C^n, \ldots.$$

Figure 2.1 shows the linear feedback shift register (LFSR) implementation of an rth-order linear recurring sequence in $GF(q)$.

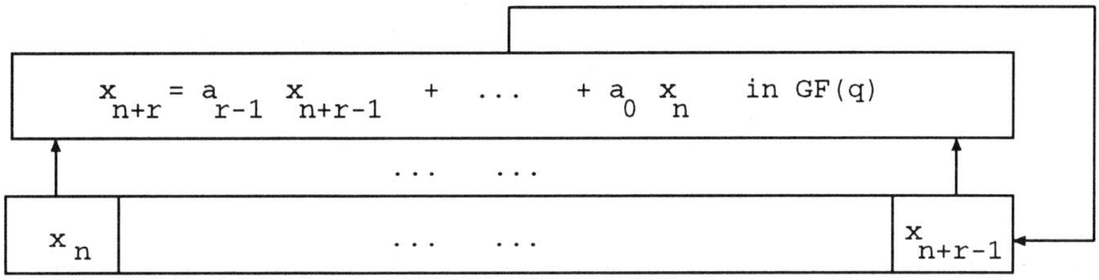

Figure 2.1 A linear feedback shift register (LFSR).

2.2 GEOMETRY OF NUMBERS

The name "geometry of numbers" was coined by Minkowski almost 100 years ago for what was then a new branch of mathematics involving both number theory and geometry. Though many results and concepts in this area are due to Minkowski himself, important later contributions were made by many other mathematicians. This section presents only the definitions of basic ideas and related fundamental results that play a crucial role in allowing us to evaluate linear congruential generators. All results are given without proofs. For detailed information, for example, see Gruber and Lekkerkerker [31].

2.2.1 Lattice

An (integer) lattice $\mathcal{L}_k$ in the k-dimensional Euclidean space $\mathbb{R}^k$ is the set of all linear combinations with integer coefficients of k linearly independent vectors in $\mathbb{R}^k$. These vectors are called a basis of the lattice $\mathcal{L}_k$. Notice that there are (countably) many sets of bases of $\mathcal{L}_k$. Figure 2.2 shows an example of a two-dimensional lattice, where not only (**a**,**b**) but also (**a**,**b**') is a lattice basis. The parallelotope generated by the basis vectors is called a fundamental parallelotope of $\mathcal{L}_k$, whose volume is equal to the determinant, denoted by $d(\mathcal{L}_k)$, of the matrix consisting of the basis vectors. The vectors of $\mathcal{L}_k$ are called lattice points. Note that the origin is a lattice point.

The Minkowski fundamental theorem is given below. Here a convex body C in $\mathbb{R}^k$ means a compact convex subset of $\mathbb{R}^k$ with a non-empty interior. $V(C)$ denotes the Lebesgue measure, namely the volume of C. We say a body is *o-symmetric* (or sometimes *balanced*) if it is symmetric with respect to the origin.

Theorem 2.28 *Let C be an o-symmetric convex body, and let $\mathcal{L}_k$ be a lattice in $\mathbb{R}^k$. If $V(C) \geq 2^k d(\mathcal{L}_k)$, then C contains at least one pair of lattice points other than the origin.*

We now give an important notion of successive minima, which will enable us to describe a lattice more quantitatively. We call a convex body *a star body* if it contains the origin as an interior point.

DEFINITION **2.16** Let C be a star body and let $\mathcal{L}_k$ be a lattice in $\mathbb{R}^k$. The successive minima $\lambda_i = \lambda_i(C, \mathcal{L}_k), i = 1, ..., k$, of C with respect to $\mathcal{L}_k$ are

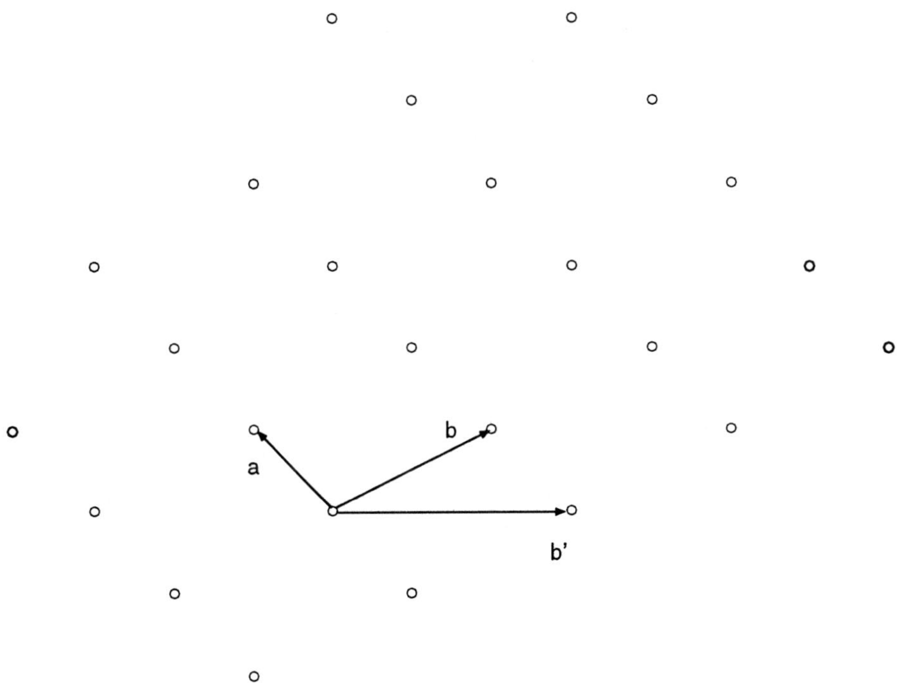

Figure 2.2 Two-dimensional integer lattice and its basis.

defined by

$$\lambda_i = \inf\{\lambda > 0 \mid \lambda C \text{ contains } i \text{ linearly independent points of } \mathcal{L}_k\}.$$

□

As an immediate consequence we have

$$0 < \lambda_1 \leq \cdots \leq \lambda_k.$$

In accordance with this notion of successive minima, the Minkowski fundamental theorem can be written in the following way:

Theorem 2.29 *Let C be an o-symmetric convex body and let $\mathcal{L}_k$ be a lattice in $\mathbb{R}^k$. Then*

$$\lambda_1^k V(C) \leq 2^k d(\mathcal{L}_k).$$

The next result is called Minkowski's second theorem.

Theorem 2.30 *Let C be an o-symmetric convex body and let $\mathcal{L}_k$ be a lattice in $\mathbb{R}^k$. Then*

$$\lambda_1 \cdots \lambda_k V(C) \leq 2^k d(\mathcal{L}_k).$$

This result is a refinement of the fundamental theorem, because $\lambda_1^k \leq \lambda_1 \cdots \lambda_k$. In the case of o-symmetric ellipsoids C, we have

$$\lambda_1 \cdots \lambda_k V(C) \leq \gamma_k^k d(\mathcal{L}_k),$$

where the constants $\gamma_k, 1 \leq k \leq 8$, are called Hermite's constants, specifically, $\gamma_k^{2k} = 1, \frac{4}{3}, 2, 4, 8, \frac{64}{3}, 64, 256$ for $k = 1, 2, ..., 8$, and are known as best possible; otherwise $\gamma_k^{2k} = (\frac{2}{\pi})^k \Gamma(k/2+2)^2$ due to Blichfeldt (see [76]), where the Gamma function $\Gamma(x)$ is defined by $\Gamma(x+1) = x\Gamma(x)$ with $\Gamma(0) = \infty$, $\Gamma(1) = 1$, and $\Gamma(\frac{1}{2}) = \sqrt{\pi}$.

As a complementary result of the above theorem, Minkowski also obtained

Theorem 2.31 *For any o-symmetric convex body C and any lattice $\mathcal{L}_k$ in $\mathbb{R}^k$,*

$$\frac{2^k}{k!} d(\mathcal{L}_k) \leq \lambda_1 \cdots \lambda_k V(C),$$

where the equality holds only if C is a cross polytope defined by $\{\mathbf{x} = (x_1, ..., x_k) \mid \sum_{i=1}^k |x_i| \leq 1\}$.

The notion of the *polar convex body* is very important. Let C be a star body in $\mathbb{R}^k$. The polar body C^* of C is defined by

$$C^* = \{\mathbf{x} \mid \langle \mathbf{x}, \mathbf{y} \rangle \leq 1 \text{ for all } \mathbf{y} \in C\},$$

where $\langle \mathbf{x}, \mathbf{y} \rangle$ means the Euclidean inner product of two k-dimensional vectors, $\mathbf{x}$ and $\mathbf{y}$. There are several important properties of polarity.

Property **2.7** *Let C be a star body. Then*

- *C^* is a star body.*
- *$C^{**} = C$.*
- *$C \subset D$ implies that $C^* \supset D^*$ if D is a star body.*
- *If C is a polytope, so is C^*.*

We have the following theorem:

Theorem 2.32 *Let C be an o-symmetric convex body. Then*

$$\frac{\kappa_k^2}{k^{k/2}} \leq V(C)V(C^*) \leq \kappa_k^2,$$

where $\kappa_k = \pi^{k/2}/\Gamma(k/2+1)$ is the volume of the k-dimensional unit ball.

The dual (or polar) lattice $\mathcal{L}_k^*$ of a lattice $\mathcal{L}_k$ is now defined by

$$\mathcal{L}_k^* = \{\mathbf{x} \mid \langle \mathbf{x}, \mathbf{y} \rangle \text{ is an integer for all } \mathbf{y} \in \mathcal{L}_k\}.$$

Let B be the $k \times k$ matrix consisting of the basis of $\mathcal{L}_k$. Then $d(\mathcal{L}_k) = |\det(B)|$. Since the lattice point of $\mathcal{L}_k$ is a linear combination of the column vectors of B, the lattice point of $\mathcal{L}_k^*$ is a linear combination of the column vectors of $B^* = ({}^tB)^{-1}$ because $B^{*t}B = I$. Note that $d(\mathcal{L}_k^*) = 1/d(\mathcal{L}_k)$. We have a famous theorem due to Mahler.

Theorem 2.33 *Let $\lambda_1, ..., \lambda_k$ be the successive minima of an o-symmetric convex body C w.r.t. $\mathcal{L}_k$, and let $\lambda_1^*, ..., \lambda_k^*$ be the successive minima of the polar convex body C^* w.r.t. $\mathcal{L}_k^*$. We then have*

$$1 \leq \lambda_i \lambda_{k+1-i}^* \leq (k!)^2$$

for all $i = 1, 2, ..., k$.

According to [31, p.134], for an o-symmetric ellipsoid C, Lenstra and Schnorr obtained an improved bound

$$1 \leq \lambda_i \lambda^*_{k+1-i} \leq k^2$$

for all $i = 1, 2, ..., k$.

The original definition of successive minima depends on a star body C as well as a lattice $\mathcal{L}_k$. In this monograph, we restrict ourselves to a narrow version of the definition by assuming that C, and thus also C^*, is a unit ball, for which the successive minima can be redefined as follows:

DEFINITION **2.17** The first successive minimum, λ_1, is the length[2] of the nonzero shortest vector $\mathbf{b}_1$ of $\mathcal{L}_k$, and for $i > 1$, the ith successive minimum λ_i is the length of the shortest vector $\mathbf{b}_i$ of $\mathcal{L}_k$ that is linearly independent of $\mathbf{b}_1, ..., \mathbf{b}_{i-1}$ over $\mathbb{R}$. □

The successive minima are an important set of constants in characterizing lattices. In general, however, the vectors corresponding to the successive minima do not form a basis for the lattice, as shown below.

EXAMPLE **2.1** *Consider a five-dimensional lattice. Let* $\mathbf{b}_1 = (1,0,0,0,0)$, $\mathbf{b}_2 = (0,1,0,\ 0,0)$, $\mathbf{b}_3 = (0,0,1,0,0)$, $\mathbf{b}_4 = (0,0,0,0,1,0)$, *and* $\mathbf{b}_5 = \frac{1}{2}(1,1,1,1,1)$. *Let* $\mathcal{L}_k$ *be a lattice whose bases are* $\mathbf{b}_i, i = 1, ..., 5$. *Note that* $|\mathbf{b}_5| = \frac{\sqrt{5}}{2} > 1$. *The successive minima of* $\mathcal{L}_k$ *are calculated as* $\lambda_1 = \cdots = \lambda_5 = 1$, *which correspond to the vectors* $\mathbf{b}_1, \mathbf{b}_2, \mathbf{b}_3, \mathbf{b}_4$, *and* $(0,0,0,0,1) = 2\mathbf{b}_5 - (\mathbf{b}_1 + \mathbf{b}_2 + \mathbf{b}_3 + \mathbf{b}_4)$. *It is easy to confirm that these vectors do not form a basis of* $\mathcal{L}_k$.

For this reason, the notion of a reduced basis of a lattice was developed as an alternative to successive minima. There are two typical kinds of reduced basis. The first one, called a Minkowski reduced basis, is defined below. Before giving its definition, we must briefly introduce some ideas from the field of quadratic forms.

[2] In this monograph, for a k-dimensional vector $\mathbf{b} = (b_1, ..., b_k)$ over $\mathbb{R}$, the length $|\mathbf{b}|$ denotes the L_2-norm $\sqrt{b_1^2 + \cdots b_k^2}$.

DEFINITION **2.18** Let $A = (a_{ij})$ be a real symmetric $k \times k$ matrix. The quadratic form $q(\mathbf{x})$ is defined by

$$q(\mathbf{x}) = {}^t\mathbf{x}A\mathbf{x} = \sum_{i,j=1}^{k} a_{ij}x_ix_j,$$

where $\mathbf{x} = {}^t(x_1, ..., x_k) \in \mathbb{R}^k$. □

We call $q(\mathbf{x})$ positive definite if $q(\mathbf{x}) > 0$ for $\mathbf{x} \neq 0$. Let T be a $k \times k$ integer matrix whose determinant is 1 or -1, that is, an integer unimodular matrix, whose geometrical interpretation is a rotation. If there exists an integer unimodular matrix T such that two quadratic forms $q_1(\mathbf{x})$ and $q_2(\mathbf{x})$ satisfy

$$q_1(\mathbf{x}) = q_2(T\mathbf{x}) \text{ for all } \mathbf{x} \in \mathbb{R}^k,$$

then we say that $q_1(\mathbf{x})$ and $q_2(\mathbf{x})$ are equivalent. Thus, we would like to have a reduced form for all equivalent quadratic forms. Minkowski defined the reduced form, in general, as follows:

DEFINITION **2.19** $q(\mathbf{x})$ is called reduced in Minkowski's sense if for each $i = 1, ..., k$ we have

$$q(x_1, ..., x_k) \geq a_{ii} = q(\overbrace{0, ..., 0}^{i-1}, 1, 0, ..., 0)$$

for any $\mathbf{x} = {}^t(x_1, ..., x_k) \in \mathbb{Z}^k$ with $\gcd(x_i, ..., x_k) = 1$. □

Note that this implies $a_{11} \leq a_{22} \leq ... \leq a_{kk}$.

There is an important relation between positive definite quadratic forms and lattices. Let $\mathcal{L}_k$ be a lattice whose basis is $\mathbf{b}_1, ..., \mathbf{b}_k$, and let B be a $k \times k$ matrix with column vectors $\mathbf{b}_1, ..., \mathbf{b}_k$. For the lattice $\mathcal{L}_k$, we define the quadratic form $q(\mathbf{x})$ as follows:

$$q(\mathbf{x}) = |B\mathbf{x}|^2 = \langle B\mathbf{x}, B\mathbf{x}\rangle = {}^t\mathbf{x}\,{}^tBB\mathbf{x} = {}^t\mathbf{x}A\mathbf{x},$$

where $A = (a_{ij}) = (\langle \mathbf{b}_i, \mathbf{b}_j\rangle) = {}^tBB$ is a symmetric matrix. It is known that for any positive definite quadratic form $q(\mathbf{x})$, there exists a lattice in the Euclidean space $\mathbb{R}^k$ whose quadratic form is $q(\mathbf{x})$. Definition 2.19 can be rewritten as follows:.

DEFINITION **2.20** A basis $\mathbf{b}_1, ..., \mathbf{b}_k$ of a lattice $\mathcal{L}_k$ is reduced in Minkowski's sense if $\mathbf{b}_1$ is the nonzero shortest vector of $\mathcal{L}_k$ and if, moreover, for $i = 2, ..., k$, $\mathbf{b}_i$ is the shortest vector over all vectors $\mathbf{b}$ of $\mathcal{L}_k$ such that $\mathbf{b}_1, ..., \mathbf{b}_{i-1}, \mathbf{b}$ can be extended to a basis of $\mathcal{L}_k$. □

We should note that the Minkowski reduced basis is not unique. The following example is due to Pohst and Zassenhaus [76]:

EXAMPLE **2.2** *Consider the hexagonal lattice in* $\mathbb{R}^3$ *whose bases are* $\mathbf{x} = (1, 0, 0)$, $\mathbf{y} = (\frac{1}{2}, \frac{\sqrt{3}}{2}, 0)$, *and* $\mathbf{z} = (\frac{1}{2}, \frac{\sqrt{3}}{6}, \frac{\sqrt{6}}{3})$. *Note that* $|\mathbf{x}| = |\mathbf{y}| = |\mathbf{z}| = 1$ *and* $\langle \mathbf{x}, \mathbf{y} \rangle = \langle \mathbf{x}, \mathbf{z} \rangle = \langle \mathbf{y}, \mathbf{z} \rangle = \frac{1}{2}$. *Thus the bases are Minkowski reduced. However, the basis* $(\mathbf{x}, \mathbf{y} - \mathbf{z}, \mathbf{z})$ *is another Minkowski reduced basis that is not isomorphic to the above, because* $\langle \mathbf{x}, \mathbf{y} - \mathbf{z} \rangle = 0$ *and* $\langle \mathbf{x}, \mathbf{z} \rangle = \langle \mathbf{z} - \mathbf{y}, \mathbf{z} \rangle = \frac{1}{2}$.

The next result describes an important connection between a Minkowski reduced basis and successive minima.

Theorem 2.34 *Let* $\mathbf{b}_i, i = 1, ..., k$, *be the Minkowski reduced basis of a* k-*dimensional lattice, and let* $\lambda_1, ..., \lambda_k$ *be its successive minima. Then we have* $|\mathbf{b}_i| = \lambda_i, 1 \le i \le \min(4, k)$, *and* $\lambda_i \le |\mathbf{b}_i| \le (\frac{5}{4})^{(i-4)/2} \lambda_i$ *for* $\min(4, k) < i \le k$.

Another popular definition of a reduced basis was proposed by Korkin and Zolotarev. We start with the Gram-Schmidt orthogonalization. Let $\mathbf{b}_1, ..., \mathbf{b}_k$ be a basis of a lattice $\mathcal{L}_k$ with $|\mathbf{b}_1| \le \cdots \le |\mathbf{b}_k|$, and let $\hat{\mathbf{b}}_1, ..., \hat{\mathbf{b}}_k$ be the orthogonal basis of the vector space spanned by $\mathcal{L}_k$ obtained by the Gram-Schmidt orthogonalization, that is,

$$\mathbf{b}_i := \sum_{j=1}^{i} \mu_{ij} \hat{\mathbf{b}}_j,$$

for $i = 1, ..., n$, where $\mu_{ij} = \left\langle \mathbf{b}_i, \hat{\mathbf{b}}_j \right\rangle / |\hat{\mathbf{b}}_j|^2$ is called the Gram-Schmidt coefficient. Assume that $\hat{\mathbf{b}}_1 = \mathbf{b}_1$ and that $\mu_{ii} = 1$ and $\mu_{ij} = 0$ for $i < j$. We call a basis $\mathbf{b}_1, ..., \mathbf{b}_k$ *proper* if $|\mu_{ij}| \le \frac{1}{2}$ for $1 \le j < i \le k$. Now, we define the Korkin-Zolotarev reduced basis:

DEFINITION **2.21** A basis $\mathbf{b}_1, ..., \mathbf{b}_k$ of a lattice $\mathcal{L}_k$ is reduced in the sense of Korkin-Zolotarev if it is a proper basis, and if $\mathbf{b}_1$ is the nonzero shortest vector of $\mathcal{L}_k$ and if, moreover, for $i = 2, ..., k$, $\mathbf{b}_i$ is the vector of $\mathcal{L}_k$ whose component orthogonal to the vector space spanned by $\{\mathbf{b}_1, ..., \mathbf{b}_{i-1}\}$ has the shortest (nonzero) length while maintaining the property that $\{\mathbf{b}_1, ..., \mathbf{b}_i\}$ can be extended to a basis of $\mathcal{L}_k$. □

In Section 2.2.2, by extending this basis, we define the Lenstra, Lenstra, and Lovász (LLL) reduced basis, which is known to be practically useful for finding short vectors of a lattice. We have an important relation between successive minima and a Korkin-Zolotarev reduced basis.

Theorem 2.35 *Every Korkin-Zolotarev reduced basis* $\mathbf{b}_1, ..., \mathbf{b}_k$ *satisfies*

$$\frac{4}{i+3} \leq \frac{|\mathbf{b}_i|^2}{\lambda_i^2} \leq \frac{i+3}{4}$$

for $i = 1, ..., k$.

Let A be a $k \times k$ matrix whose column vectors are $\mathbf{a}_1, ..., \mathbf{a}_k$. Hadamard's inequality reads

$$|\det(A)| \leq \prod_{i=1}^{k} |\mathbf{a}_i|.$$

The equality holds only if $\mathbf{a}_i, i = 1, ..., k$, are all orthogonal. Note that the Gram-Schmidt orthogonal bases of a given lattice are all orthogonal. We now define the orthogonality defect of a lattice as follows:

DEFINITION **2.22** Let B be a $k \times k$ matrix whose column vectors $\mathbf{b}_i, i = 1, ..., k$, are a reduced basis of a lattice $\mathcal{L}_k$. The orthogonality defect of $\mathcal{L}_k$ is defined by

$$\mathrm{OD}(\mathbf{b}_1, ..., \mathbf{b}_k) \stackrel{\mathrm{def}}{=} \frac{\prod_{i=1}^{k} |\mathbf{b}_i|}{|\det(B)|}.$$

□

Note that $\mathrm{OD}(\mathbf{b}_1, ..., \mathbf{b}_k) \geq 1$ for any basis $\mathbf{b}_1, ..., \mathbf{b}_k$. For the Korkin-Zolotarev basis, the orthogonality defect is known to be at most k^k, which is much smaller than $2^{O(k^2)}$ for that of the Minkowski reduced basis.

Polynomial Lattice

Next, we give several definitions and related results for the analogue of the geometry of numbers in $GF\{q,z\}^k$, originated by Mahler [53]. We consider a polynomial lattice $\mathcal{L}_k$ in $GF\{q,z\}^k$, namely, the set of all linear combinations with polynomial coefficients of k linearly independent vectors in $GF\{q,z\}^k$. These vectors are called a basis of the lattice $\mathcal{L}_k$, and there are (countably) many sets of bases of $\mathcal{L}_k$. The vectors of $\mathcal{L}_k$ are called lattice points. Note that the origin is a lattice point. Let each $f_i(z), i = 1, ..., k$, be in $GF\{q,z\}$. Define the norm of a vector $\mathbf{f} = (f_1(z), ..., f_k(z))$ as

$$|\mathbf{f}| = \max_{1 \le i \le k} \nu(f_i),$$

where ν is the discrete exponential valuation defined in Section 2.1.

The original definition of successive minima for a polynomial lattice depends on a star body C as well as a lattice $\mathcal{L}_k$. However, in this monograph, we use a restricted version of the definition by assuming that C is a set of points X such that $|X| < 0$.

DEFINITION **2.23** The first successive minimum λ_1 is defined as the norm of a nonzero shortest vector $\mathbf{b}_1$ of a lattice $\mathcal{L}_k$ in $GF\{q,z\}^k$. For $2 \le i \le k$, a ith successive minimum λ_i of $\mathcal{L}_k$ is recursively defined as the norm of a smallest vector $\mathbf{b}_i$ in $\mathcal{L}_k$ that is linearly independent of $\mathbf{b}_1, \mathbf{b}_2, ..., \mathbf{b}_{i-1}$ over $GF\{q,z\}$. □

As an immediate consequence, we have

$$-\infty < \lambda_1 \le \cdots \le \lambda_k.$$

The following result is important [54, Lemma 1]:

Theorem 2.36 *A set of vectors,* $\mathbf{b}_1, ..., \mathbf{b}_k$*, corresponding to the successive minima* $\lambda_1, ..., \lambda_k$ *of a lattice* $\mathcal{L}_k$*, forms a basis of* $\mathcal{L}_k$ *over* $GF\{q,z\}$*.*

Since any cube $C_l^{(k)} = \{X \mid |X| < l\}$, $l \in \mathbb{Z}$, contains a finite number of points in a given lattice $\mathcal{L}_k$, it follows that a basis in the above theorem always exists. This basis is called a *reduced* basis for $\mathcal{L}_k$.

The analogue of Minkowski's fundamental theorem is as follows: Let B be the $k \times k$ matrix whose column vectors $\mathbf{b}_1, ..., \mathbf{b}_k$ are the basis of $\mathcal{L}_k$. Define $d(\mathcal{L}_k) = |\det(B)|$.

Theorem 2.37 *We have*

$$k\lambda_1 \le d(\mathcal{L}_k).$$

The next result is the analogue of Minkowski's second theorem.

Theorem 2.38 *We have*

$$\lambda_1 + \cdots + \lambda_k = d(\mathcal{L}_k).$$

The dual (or polar) lattice $\mathcal{L}_k^*$ of a lattice $\mathcal{L}_k$ is defined by

$$\mathcal{L}_k^* = \{\mathbf{x} \mid \langle \mathbf{x}, \mathbf{y} \rangle \text{ is a polynomial for all } \mathbf{y} \in \mathcal{L}_k\}.$$

Since the lattice point of $\mathcal{L}_k$ is a linear combination of the column vectors of B, the lattice point of $\mathcal{L}_k^*$ is a linear combination of the column vectors of $B^* = ({}^tB)^{-1}$ because $B^{*t}B = I$. Note that $d(\mathcal{L}_k^*) + d(\mathcal{L}_k) = 0$. We have a famous theorem due to Mahler [53]:

Theorem 2.39 *Let $\lambda_1, ..., \lambda_k$ be the successive minima of a lattice $\mathcal{L}_k$, and let $\lambda_1^*, ..., \lambda_k^*$ be the successive minima of the dual lattice $\mathcal{L}_k^*$. We then have*

$$\lambda_i + \lambda_{k+1-i}^* = 0$$

for all $i = 1, 2, ..., k$.

The orthogonality defect is defined by

$$\mathrm{OD}(\mathbf{b}_1, ..., \mathbf{b}_k) = \sum_{i=1}^{k} |\mathbf{b}_i| - |\det(B)|.$$

Lenstra [47] obtained

Theorem 2.40 *Let $\mathbf{b}_1, ..., \mathbf{b}_k$ be a basis of $\mathcal{L}_k$ such that $\mathrm{OD}(\mathbf{b}_1, ..., \mathbf{b}_k) = 0$ with $|\mathbf{b}_1| \le \cdots \le |\mathbf{b}_k|$. Then $|\mathbf{b}_i|$ is an ith successive minimum of $\mathcal{L}_k$ for $i = 1, ..., k$.*

Therefore, the analogous versions of Korkin-Zolotarev and Minkowski reduced bases for polynomial lattices are equivalent.

2.2.2 Basis Reduction Algorithms

We describe several algorithms relevant to basis reduction of a given lattice basis. Before that, we need to describe an important subroutine SOLVEQX given in Figure 2.3, which finds all $\mathbf{x} \in \mathbb{Z}^k$ satisfying the quadratic equation $q(\mathbf{x}) \equiv \mathbf{x}^t A\mathbf{x} \leq c$, where c is a positive constant. This algorithm, due to Fincke and Pohst [29], is based on Cholesky decomposition of the positive definite matrix A, to get

$$q(\mathbf{x}) = \sum_{i=1}^{k} q_{ii} \left(x_i + \sum_{j=i+1}^{k} q_{ij} x_j \right)^2 .$$

The algorithm can be applied to the following problems:

1. Computation of the shortest nonzero vector in a lattice
2. Enumeration of all $\mathbf{x} \in \mathbb{Z}^k$ satisfying $0 < c' \leq {}^t\mathbf{x}A\mathbf{x} \leq c$
3. Computation of the closest point in a lattice to a given vector

The algorithm RBASIS, due to Pohst and Zassenhaus [76], finds a Minkowski reduced basis for a given lattice, where SOLVEQX is used as a subroutine to solve Problem 2 in the above list. First, we need to define a partial ordering of vectors in $\mathbb{R}^k$ in the following way:

$$\mathbf{a} < \mathbf{b} \Leftrightarrow |\mathbf{a}| < |\mathbf{b}|$$

for $\mathbf{a}, \mathbf{b} \in \mathbb{R}^k$. By using this ordering, we can order all the bases of a k-dimensional lattice over $\mathbb{R}$ as follows:

$$(\mathbf{b}_1, ..., \mathbf{b}_k) < (\mathbf{a}_1, ..., \mathbf{a}_k)$$

if and only if there exists an integer j such that $|\mathbf{b}_i| = |\mathbf{a}_i|$ for any $0 \leq i \leq j-1 < k$ and $\mathbf{b}_j < \mathbf{a}_j$, where we assume $|\mathbf{a}_0| = |\mathbf{b}_0| = 0$. The algorithm SUCC for computing successive minima for a given lattice is described in Figure 2.5, which is slightly simpler than RBASIS. SOLVEQX is also used as a subroutine in the algorithm SUCC, which needs to solve Problem 3 in the list. The integer unimodular matrix U needed in the algorithms RBASIS and SUCC can be easily constructed by using the following lemma [76, Lemma 2.10]:

Lemma 2.1 *Let* $\xi_1, ..., \xi_k$ *be integers with* $\gcd(\xi_1, ..., \xi_r) = d_r$ *for* $r = 1, 2, ..., k$. *Note that* $d_r = \gcd(d_{r-1}, \xi_r)$ *with* $d_1 = \xi_1$. *Let* $d_r = u_r d_{r-1} + v_r \xi_r$ *and* $m_{11} =$

Procedure SOLVEQX
input: A positive definite matrix A and a constant $c > 0$
output: All nonzero $\mathbf{x} = (x_1, ..., x_k) \in \mathbb{Z}^k$ satisfying $q(\mathbf{x}) \equiv \mathbf{x}^t A\mathbf{x} \le c$
begin
  Set $q_{ij} = a_{ij}$ $(1 \le i \le j \le k)$;
  for $i = 1, 2, ..., k-1$ **do**
    for $j = i+1, ..., k$ **do**
      $q_{ji} = q_{ij}$; $q_{ij} = q_{ij}/q_{ii}$;
    end
    for $m = i+1, ..., k$ **do**
      for $l = m, ..., k$ **do**
        $q_{ml} = q_{ml} - q_{mi}q_{il}$;
      end
    end
  end
  Set $q_{ij} = 0$ $(1 \le j < i \le k)$;
  Set $i = k$, $T_i = c$, and $U_i = 0$;
  `label0`: $Z = \sqrt{T_i/q_{ii}}$; $UB(x_i) = \lfloor Z - U_i \rfloor$; $x_i = \lceil -Z - U_i \rceil - 1$;
  `label1`: $x_i = x_i + 1$;
  if $x_i > UB(x_i)$ **then**
    $i = i + 1$; **goto** `label1`;
  if $i > 1$ **then**
    $i = i - 1$; $U_i = \sum_{j=i+1}^{k} q_{ij}x_j$; $T_i = T_{i+1} - q_{i+1,i+1}(x_{i+1} + U_{i+1})^2$;
    goto `label0`;
  else
    if $\mathbf{x} \ne \mathbf{0}$ **then**
      print $\mathbf{x}$, $-\mathbf{x}$, and $q(\mathbf{x}) = c - T_1 + q_{11}(x_1 + U_1)^2$; **goto** `label1`;
end

Figure 2.3 Algorithm SOLVEQX.

ξ_1/d_k. *For* $r = 2, ..., k$, *if we recursively define* $m_{rr} = u_r$, $m_{rj} = -(\xi_j/d_{r-1})v_r$ $(1 \le j \le r-1)$, $m_{1r} = \xi_r/d_k$, *and* $m_{ir} = 0$ $(2 \le i \le r-1)$, *then the matrix* $U = (m_{ij})$ *becomes a* $k \times k$ *integer unimodular matrix satisfying* $(d_k, 0, ..., 0)U = (\xi_1, ..., \xi_k)$.

Next, we define the LLL reduced basis [48], the modified version of the Korkin-Zolotarev reduced basis, and describe the algorithm LLLBASIS, which we recommend the reader to use as a preprocessor for the algorithms RBASIS and SUCC, because it efficiently provides a basis very close to the Korkin-Zolotarev basis. Fincke and Pohst [29] reported that incorporating the LLL basis reduction into the Cholesky decomposition will further speed up the algorithm SOLVEQX, particularly in high dimensions.

DEFINITION **2.24** A basis $\mathbf{b}_1, ..., \mathbf{b}_k$ is called LLL reduced with δ, $\frac{1}{4} < \delta < 1$, if it is proper and if

$$\delta|\hat{\mathbf{b}}_i|^2 \le |\hat{\mathbf{b}}_{i+1} + \mu_{i+1,i}\hat{\mathbf{b}}_i|^2$$

for $i = 1, 2, ..., k-1$, where $\hat{\mathbf{b}}_1, ..., \hat{\mathbf{b}}_k$ is the Gram-Schmidt orthogonalization. □

The properties of the LLL reduced basis are summarized as follows:

Property **2.8** *Let* $\mathbf{b}_1, ..., \mathbf{b}_k$ *be the LLL reduced basis.*

- $|\mathbf{b}_1| \le 2^{(k-1)/4} \det(\mathcal{L}_k)^{1/k}$;
- $|\mathbf{b}_1| \le 2^{(k-1)/2} \min\{|\mathbf{x}| : \mathbf{x} \in \mathcal{L}_k, \mathbf{x} \ne 0\}$;
- $\det(\mathcal{L}_k) \le |\mathbf{b}_1| \cdots |\mathbf{b}_k| \le 2^{k(k-1)/4} \det(\mathcal{L}_k)$;

The last property means that for the LLL basis,

$$\mathrm{OD}(\mathbf{b}_1, ..., \mathbf{b}_k) \le 2^{k(k-1)/4}.$$

We have a relation between successive minima and the LLL basis [48].

Theorem 2.41 *Every LLL reduced basis* $\mathbf{b}_1, ..., \mathbf{b}_k$ *with* δ *satisfies*

$$\alpha^{1-i} \le \frac{|\mathbf{b}_i|^2}{\lambda_i^2} \le \alpha^{k-1}$$

for $i = 1, ..., k$, *where* $\alpha = (\delta - \frac{1}{4})^{-1}$.

```
Procedure RBASIS
input:  b_1, ..., b_k a basis of rank k ≥ 2
output: b_1, ..., b_k a Minkowski reduced basis
begin
   r = 1; Λ_r = Zb_1; \hat{b}_1 = b_1;
   while r < k do
      Compute μ_{r+1,j} = ⟨b_{r+1}, \hat{b}_j⟩ / |\hat{b}_j|^2 for 1 ≤ j ≤ r;
      Set \hat{b}_{r+1} = b_{r+1} − Σ_{j=1}^{r} μ_{r+1,j} \hat{b}_j;
      Compute β_{r+1,j} such that \hat{b}_{r+1} = b_{r+1} − Σ_{j=1}^{r} β_{r+1,j} b_j;
      Set B_{r+1} = ⌊|b_{r+1}|/|\hat{b}_{r+1}|⌋ and ξ_{r+1} = 1;
      while ξ_{r+1} ≤ B_{r+1} do
         Set B(ξ_{r+1}) = |b_{r+1}|^2 − ξ_{r+1}^2 |\hat{b}_{r+1}|^2;
         Call SOLVEQX to compute M defined by
         M = {x = Σ_{i=1}^{r+1} ξ_i b_i ∈ Λ_r + Zb_{r+1} | |x − ξ_{r+1} \hat{b}_{r+1}|^2 ≤ B(ξ_{r+1})};
         if M ≠ φ then
            for j = 1, ..., r + 1 do
               Let x_j be minimal in M such that gcd(ξ_j, ..., ξ_{r+1}) = 1 and x_j < b_j;
               if no such x_j exists then x_j = 0;
            end
            if x_j ≠ 0 exists for some j (1 ≤ j ≤ r + 1) then
               for each j with x_j ≠ 0 do
                  Let U be a unimodular matrix of size r + 2 − j over integers,
                  whose first column is (ξ_j, ..., ξ_{r+1});
                  Transform the bases b_j, ..., b_{r+1} into the new ones b'_j = x_j, b'_{j+1},
                  ..., b'_{r+1}, by using (b'_j, ..., b'_{r+1}) = (b_j, ..., b_{r+1})U;
                  Update \hat{b}_i (j ≤ i ≤ r + 1) and the coordinates of x_i (i > j);
               end
               Let j_0 be the smallest index for which a transformation occurred;
               if j_0 ≤ r then r = j_0; Λ_r = Zb_1 + ··· + Zb_r; goto label0;
            else ξ_{r+1} = ξ_{r+1} + 1;
      end
      Λ_{r+1} = Λ_r + Zb_{r+1}; r = r + 1;
   label0: end
end
```

Figure 2.4 Algorithm RBASIS.

Procedure SUCC
input: $\mathbf{b}_1, ..., \mathbf{b}_k$ a basis of rank $k \geq 2$
output: $|\mathbf{y}_1|, ..., |\mathbf{y}_k|$ the successive minima
begin
  $r = 1; \Lambda_r = \mathbb{Z}\mathbf{b}_1; \hat{\mathbf{b}}_1 = \mathbf{b}_1; \mathbf{y}_j = \mathbf{b}_j \ (1 \leq j \leq k)$;
  while $r < k$ **do**
    Compute $\mu_{r+1,j} = \langle \mathbf{b}_{r+1}, \hat{\mathbf{b}}_j \rangle / |\hat{\mathbf{b}}_j|^2$ for $1 \leq j \leq r$;
    Set $\hat{\mathbf{b}}_{r+1} = \mathbf{b}_{r+1} - \sum_{j=1}^{r} \mu_{r+1,j} \hat{\mathbf{b}}_j$;
    Compute $\beta_{r+1,j}$ such that $\hat{\mathbf{b}}_{r+1} = \mathbf{b}_{r+1} - \sum_{j=1}^{r} \beta_{r+1,j} \mathbf{b}_j$;
    Set $B_{r+1} = \lfloor |\mathbf{b}_{r+1}| / |\hat{\mathbf{b}}_{r+1}| \rfloor$ and $\xi_{r+1} = 1$;
    while $\xi_{r+1} \leq B_{r+1}$ **do**
      Call **SOLVEQX** to compute $\mathbf{b} \in \Lambda_r + \xi_{r+1}(\mathbf{b}_{r+1} - \hat{\mathbf{b}}_{r+1})$ of shortest length;
      Set $\mathbf{x} = \mathbf{b} + \xi_{r+1} \hat{\mathbf{b}}_{r+1}$;
      Compute $m_i \ (1 \leq i \leq r+1)$ such that $\mathbf{x} = \sum_{i=1}^{r+1} m_i \mathbf{b}_i$;
      if $|\mathbf{x}| < |\mathbf{y}_{r+1}|$ **then**
        Let $j \in \{1, ..., r+1\}$ be minimal with $|\mathbf{x}| < |\mathbf{y}_j|$;
        Set $a = \gcd(m_j, ..., m_{r+1})$;
        Let U be a unimodular matrix of size $r + 2 - j$ over integers,
        whose first column is $(m_j/a, ..., m_{r+1}/a)$;
        Transform the bases $\mathbf{b}_j, ..., \mathbf{b}_{r+1}$ into the new ones $\mathbf{b}'_j = \frac{1}{a}(\mathbf{x} - \sum_{i=1}^{j-1} m_i \mathbf{b}_i)$,
        $\mathbf{b}'_{j+1}, ..., \mathbf{b}'_{r+1}$, by using $(\mathbf{b}'_j, ..., \mathbf{b}'_{r+1}) = (\mathbf{b}_j, ..., \mathbf{b}_{r+1})U$;
        Update $\hat{\mathbf{b}}_i \ (j \leq i \leq r+1)$;
        Set $\mathbf{y}_j = \mathbf{x}$ and $r = j$;
        goto `label0`;
      else $\xi_{r+1} = \xi_{r+1} + 1$;
    end
    $\Lambda_{r+1} = \Lambda_r + \mathbb{Z}\mathbf{b}_{r+1}$; $r = r + 1$;
  `label0`: **end**
end

Figure 2.5 Algorithm SUCC.

This implies that any LLL basis consists of relatively short vectors. The relation between a Korkin-Zolotarev basis and an LLL basis is as follows: Let $b_{i,j} = \left\langle \mathbf{b}_i, \hat{\mathbf{b}}_j \right\rangle / |\hat{\mathbf{b}}_j|$. Note that $b_{i,j} = 0$ for $i < j$. From the definition of the Korkin-Zolotarev reduced basis $\mathbf{b}_1, ..., \mathbf{b}_k$, we have

$$b_{i,i}^2 \leq b_{i+1,i}^2 + b_{i+1,i+1}^2$$

for $i = 1, 2, ..., k-1$, because $|b_{i,i}|$ is the length of the shortest vector in the lattice generated by the bottom right $(k-i+1) \times (k-i+1)$ submatrix of the basis matrix. Since the basis is proper, we have $|b_{i,i}| \geq 2|b_{i+1,i}|$. Thus $\frac{3}{4}b_{i,i}^2 \leq b_{i+1,i+1}^2$, that is, $b_{i,i}^2 \leq b_{i+1,i+1}^2 + \frac{1}{4}b_{i,i}^2$, which corresponds to the case of the LLL basis with $\delta = 1$ and $\mu_{i+1,i} = \frac{1}{2}$.

In Figure 2.6, we give an algorithm, LLLBASIS, which computes the LLL basis for a given lattice basis, where $\lceil x \rfloor$ denotes the nearest integer to x.

```
Procedure LLLBASIS
input:   b_1, ..., b_k ∈ Z^k a basis of rank k ≥ 2
output:  b_1, ..., b_k a LLL basis with 1/4 < δ < 1
begin
  m = 2;
  Compute the Gram-Schmidt coefficients μ_ij (1 ≤ j < i ≤ k) and |b̂_1|^2, ..., |b̂_k|^2;
  while m ≤ k do
    for j = m − 1, ..., 1 do
      if |μ_{m,j}| > 1/2 then
        b_m = b_m − ⌈μ_{m,j}⌋ b_j;
        for i = 1, ..., j do
          μ_{m,i} = μ_{m,i} − ⌈μ_{m,j}⌋ μ_{j,i};
        end
    end
    if δ|b̂_{m−1}|^2 > |b̂_m|^2 + μ_{m,m−1}^2 |b̂_{m−1}|^2 then
      Swap b_m and b_{m−1}, where b̂_{m−1} is replaced by b̂_m + μ_{m,m−1} b̂_{m−1};
      m = max(m − 1, 2);
    else m = m + 1;
  end
end
```

Figure 2.6 Algorithm LLLBASIS.

For practical purposes, δ is taken close to 1, say, 0.999. Let $B_{max} = \max_{i=1}^{k} |\mathbf{b}_i|$ for the input basis. The data size needed for $\mathbf{b}_i$ and μ_{ij} throughout the algorithm is $O(k \log B_{max})$, which means that we need multiple-precision arithmetic. In order to overcome this, Schnorr [83] proposed a floating-point arithmetic version of the LLL algorithm.

Polynomial Lattice

In Figure 2.7, we give the basis reduction algorithm LENSTRA due to Lenstra for a lattice in $GF\{q,z\}^k$. The algorithm is based on his following result:

Proposition 2.1 *Let $\mathbf{b}_1, ..., \mathbf{b}_k$ be a basis of a lattice in $GF\{q,z\}^k$. Let $B = (b_{ij})$ be a $k \times k$ matrix with rows $\mathbf{b}_1, ..., \mathbf{b}_k$. Then the basis is reduced if the columns of B can be permuted in such a way that the rows $\bar{\mathbf{b}}_1, ..., \bar{\mathbf{b}}_k$ of the resulting matrix $\bar{B} = (\bar{b}_{ij})$ satisfy*

$$|\bar{\mathbf{b}}_i| \leq |\bar{\mathbf{b}}_j| \text{ for } 1 \leq i < j \leq k, \tag{2.4}$$

$$|\bar{b}_{ii}| \geq |\bar{b}_{ij}| \text{ for } 1 \leq i < j \leq k, \tag{2.5}$$

and

$$|\bar{b}_{ii}| > |\bar{b}_{ij}| \text{ for } 1 \leq j < i \leq k. \tag{2.6}$$

Proof: This follows from the fact that a basis $\mathbf{b}_1, ..., \mathbf{b}_k$ in the above satisfies $\text{OD}(\mathbf{b}_1, ..., \mathbf{b}_k) = 0$; that is, $|\mathbf{b}_i|$ is a ith successive minimum for $i = 1, ..., k$. □

Note that at every stage $m \in \{0, 1, ..., k\}$, the algorithm proceeds in such a way that the basis satisfies the following conditions:

$$|\mathbf{b}_i| \leq |\mathbf{b}_j| \text{ for } 1 \leq i < j \leq m,$$

$$|\mathbf{b}_m| \leq |\mathbf{b}_j| \text{ for } m < j \leq k,$$

$$|b_{ii}| \geq |b_{ij}| \text{ for } 1 \leq i \leq m \text{ and } i < j \leq k,$$

and

$$|b_{ii}| > |b_{ij}| \text{ for } 1 \leq j < i \leq m.$$

From these conditions, we have $a_{ii} \neq 0$ for $1 \leq i \leq m$, and $a_{ij} = 0$ for $1 \leq j < i \leq m$. Hence, the matrix $A = (a_{ij})$ in the algorithm is nonsingular and thus there always exists a solution for this linear system.

```
Procedure LENSTRA
input:  b_1, ..., b_k ∈ GF[q, z]^k a basis of rank k ≥ 2
output: b_1, ..., b_k a reduced basis permuted so that (2.4)–(2.6) hold
begin
  m = 0; |b_0| = −∞;
  while m < k do
    Renumber {b_{m+1}, ..., b_k} in such a way that |b_{m+1}| = min_{m+1≤i≤k} |b_i|;
    if m > 0 then
      Construct A = (a_ij)_{1≤i,j≤m} be an m × m matrix such that a_ij ∈ GF(q)
      be the coefficient of z^{|b_i|} in b_ij for 1 ≤ i ≤ m + 1 and 1 ≤ j ≤ m;
      Let x = ^t(x_1, ..., x_m) and b = ^t(a_{m+1,1}, ..., a_{m+1,m}) be in GF(q)^m;
      Solve Ax = b;
    Put b̂_{m+1} = b_{m+1} − Σ_{i=1}^{m} x_i b_i z^{|b_{m+1}|−|b_i|};
    if |b̂_{m+1}| = |b_{m+1}| then
      b_{m+1} = b̂_{m+1};
      Permute the coordinates of b_1, ..., b_k so that |b_{m+1,m+1}| = |b_{m+1}|;
      m = m + 1;
    else
      b_{m+1} = b̂_{m+1};
      m = max{l | |b_l| ≤ |b_{m+1}|, l = 0, 1, ..., m};
  end
end
```

Figure 2.7 Algorithm LENSTRA.

Lenstra also proved that the algorithm runs in $O(k^3 B_{max}(\mathrm{OD}(\mathbf{b}_1, \ldots, \mathbf{b}_k) + 1))$ time, where the unit is the arithmetic operation in $GF(q)$ and $B_{max} = \max_{i=1}^{k} |\mathbf{b}_i|$. Since $\mathrm{OD}(\mathbf{b}_1, \ldots, \mathbf{b}_k) \leq kB_{max}$, the time complexity becomes $O(k^4 B_{max}^2)$. He also remarked that the algorithm runs faster than the algorithm based on Gaussian elimination, which takes $O(k^6 B_{max}^3)$.

2.3 UNIFORM DISTRIBUTION OF SEQUENCES

The theory of uniform distribution of sequences was originated by Hermann Weyl in 1916. He based it firmly on Diophantine approximations, but nowadays it has come to exist in its own right. In this section, we briefly overview some important results in this field, which are fundamental tools for the analysis of the uniformity of several types of random numbers discussed in later chapters.

2.3.1 Weyl's Criterion

We start with the definition of the uniform distribution of sequences modulo one.

DEFINITION **2.25** Let $x_n, n = 1, 2, \ldots$, be an (infinite) sequence of real numbers. Let $\#(E; N)$ be defined as the number of elements $x_n, 1 \le n \le N$, whose fractional parts are in E, a subinterval of $[0, 1]$. We call the sequence uniformly distributed modulo one if

$$\lim_{N\to\infty} \frac{\#([a,b); N)}{N} = b - a,$$

for any real numbers a and b such that $0 \le a < b \le 1$. □

Note that only the fractional parts of $x_n, n = 1, 2, \ldots$, matter for the notion of uniform distribution modulo one. The notion of uniform distribution of sequences has an important link with the integration of a function over $[0, 1]$. Let $\{x\}$ denote the fractional part of a real number x.

Theorem 2.42 *A sequence $x_n, n = 1, 2, \ldots$, is uniformly distributed modulo one if and only if*

$$\lim_{N\to\infty} \frac{1}{N} \sum_{n=1}^{N} f(\{x_n\}) = \int_0^1 f(x)dx,$$

for any real-valued continuous function f defined on $[0, 1]$.

Next, we give Weyl's celebrated result, which is a central theorem in the classical part of the theory of uniform distribution.

Theorem 2.43 *A sequence $x_n, n = 1, 2, ...,$ is uniformly distributed modulo one if and only if for all integers $h \neq 0$*

$$\lim_{N\to\infty} \frac{1}{N} \sum_{n=1}^{N} e^{2\pi\sqrt{-1}hx_n} = 0.$$

The result is a very powerful tool in analyzing uniform distribution of many sequences.

EXAMPLE **2.3** *If θ is an irrational number, then the Weyl sequence $n\theta, n = 1, 2, ...,$ is uniformly distributed modulo one.*

A stricter notion of uniform distribution was developed as follows:

DEFINITION **2.26** Let $\#(E; N, d)$ be the number of elements among $x_{d+1}, x_{d+2}, ..., x_{d+N}$ whose fractional parts are in E. We call the sequence $x_n, n = 1, 2, ...,$ well-distributed modulo one if

$$\lim_{N\to\infty} \frac{\#([a,b); N, d)}{N} = b - a \text{ uniformly in } d = 0, 1, ...,$$

for any real numbers a and b such that $0 \leq a < b \leq 1$. □

A uniformly distributed sequence modulo one is not necessarily well-distributed modulo one. A counterexample is found in [44, Example 5.1]. The following theorems are generalizations of Theorems 2.42 and 2.43:

Theorem 2.44 *A sequence $x_n, n = 1, 2, ...,$ is well-distributed modulo one if and only if*

$$\lim_{N\to\infty} \frac{1}{N} \sum_{n=d+1}^{d+N} f(\{x_n\}) = \int_0^1 f(x)dx \text{ uniformly in } d = 0, 1, ...,$$

for any real-valued continuous function f defined on $[0, 1]$.

Theorem 2.45 *A sequence $x_n, n = 1, 2, ...,$ is well-distributed modulo one if and only if for all integers $h \neq 0$*

$$\lim_{N\to\infty} \frac{1}{N} \sum_{n=d+1}^{d+N} e^{2\pi\sqrt{-1}hx_n} = 0 \text{ uniformly in } d = 0, 1, ...$$

Note that the Weyl sequence $n\theta, n = 1, 2, ...,$ is also well-distributed modulo one. However, we can rarely find other well-distributed sequences.

The multidimensional case of uniform distribution modulo one is defined as follows:

DEFINITION **2.27** Let $\mathbf{x}_n, n = 1, 2, ...,$ be an (infinite) sequence of vectors in $\mathbb{R}^k$. Let $\#(E; N)$ be defined as the number of elements $\mathbf{x}_n, 1 \le n \le N$, whose fractional parts are in E, a subinterval of $[0, 1]^k$. We call the sequence uniformly distributed modulo one if

$$\lim_{N\to\infty} \frac{\#([\mathbf{a}, \mathbf{b}); N)}{N} = \prod_{i=1}^{k} (b_i - a_i),$$

for any k-dimensional points $\mathbf{a} = (a_1, ..., a_k)$ and $\mathbf{b} = (b_1, ..., b_k)$ such that $[\mathbf{a}, \mathbf{b}) \in [0, 1]^k$. □

Thus we have similar results to those in the one-dimensional case.

Theorem 2.46 *A sequence $\mathbf{x}_n, n = 1, 2, ...,$ is uniformly distributed modulo one in $\mathbb{R}^k$ if and only if*

$$\lim_{N\to\infty} \frac{1}{N} \sum_{n=1}^{N} f(\{\mathbf{x}_n\}) = \int_{[0,1]^k} f(\mathbf{x}) d\mathbf{x},$$

for any real-valued continuous function f defined on $[0, 1]^k$, and $\{\mathbf{x}\}$ denotes $(\{x_1\}, ..., \{x_k\})$, i.e., the coordinate-wise fractional part of $\mathbf{x} = (x_1, ..., x_k)$.

Theorem 2.47 *A sequence $\mathbf{x}_n, n = 1, 2, ...,$ is uniformly distributed modulo one in $\mathbb{R}^k$ if and only if for all lattice points $\mathbf{h} \in \mathbb{Z}^k$ with $\mathbf{h} \neq 0$,*

$$\lim_{N\to\infty} \frac{1}{N} \sum_{n=1}^{N} e^{2\pi\sqrt{-1}\langle \mathbf{h}, \mathbf{x}_n \rangle} = 0,$$

where $\langle \mathbf{x}, \mathbf{y} \rangle$ denotes the inner product of $\mathbf{x}$ and $\mathbf{y}$ in $\mathbb{R}^k$.

We have one example of uniformly distributed sequences modulo one in $\mathbb{R}^k$.

EXAMPLE **2.4** *The multidimensional version of the Weyl sequence, that is,* $(n\theta_1, ..., n\theta_k), n = 1, 2, ...,$ *is uniformly distributed modulo one in* $\mathbb{R}^k$, *if the reals* $1, \theta_1, ..., \theta_k$ *are linearly independent over the rationals.*

We now define the important notion of *k-distribution* of a sequence of real numbers, $x_n, n = 1, 2, ...,$ as follows:

DEFINITION **2.28** A sequence $x_n, n = 1, 2, ...,$ in $[0, 1)$ is said to be k-distributed if $\mathbf{x}_n = (x_n, ..., x_{n+k-1}), n = 1, 2, ...,$ is uniformly distributed modulo one in $\mathbb{R}^k$. □

DEFINITION **2.29** A sequence $x_n, n = 1, 2, ...,$ in $[0, 1)$ is said to be ∞-distributed if it is k-distributed for all $k = 1, 2, ...$ □

The next theorem guarantees the existence of ∞-distributed sequences.

Theorem 2.48 *A sequence* $x_n = \theta^n, n = 1, 2, ...,$ *is* ∞*-distributed for almost all real numbers* $\theta > 1$.

For more details about the notion of ∞-distribution, we refer the reader to Knuth's book [40], in which he goes deeply into an interesting question: "Does ∞-distributed = random?"

In what follows, we elaborate on the discrete version of the results given above.

DEFINITION **2.30** Let $a_n, n = 1, 2, ...,$ be a sequence of integers. For integers $N \geq 1$, $m \geq 2$, and j, we define $\#(j, m, N)$ as the number of elements among $a_1, ..., a_N$ satisfying $a_n \equiv j \pmod{m}$ with $1 \leq n \leq N$. We call the sequence uniformly distributed modulo m if

$$\lim_{N \to \infty} \frac{\#(j, m, N)}{N} = \frac{1}{m}$$

for all $j = 0, 1, ..., m - 1$. □

EXAMPLE **2.5** *Let* $a_n = n \pmod{m}$ *for* $n = 1, 2, ...$ *Then a (periodic) sequence* $a_n, n = 1, 2, ...,$ *is uniformly distributed modulo* k *if and only if* $k > 1$ *divides* m.

The discrete version of the Weyl criterion is as follows:

Theorem 2.49 *A sequence $a_n, n = 1, 2, ...,$ of integers is uniformly distributed modulo m if and only if for all integers $h \neq 0 \pmod{m}$,*

$$\lim_{N\to\infty} \frac{1}{N} \sum_{n=1}^{N} e^{2\pi\sqrt{-1}ha_n/m} = 0.$$

The multidimensional version of the above results can be obtained in the same way as for sequences of reals, that is,

DEFINITION **2.31** Let $\mathbf{a}_n = (a_{n,1}, ..., a_{n,k}), n = 1, 2, ...,$ be an (infinite) sequence of vectors in $\mathbb{Z}^k$. Let $\#(\mathbf{j}, m, N)$ be defined as the number of elements $\mathbf{a}_n, 1 \leq n \leq N$, such that $a_{n,i} = j_i \pmod{m}$ for all $1 \leq i \leq k$. We call the sequence uniformly distributed modulo m if

$$\lim_{N\to\infty} \frac{\#(\mathbf{j}, m, N)}{N} = \frac{1}{m^k},$$

for all $\mathbf{j} = (j_1, ..., j_k) \in \mathbb{Z}^k$ with $0 \leq j_1, ..., j_k \leq m - 1$. □

Thus we have similar results to those in the one-dimensional case.

Theorem 2.50 *A sequence $\mathbf{a}_n, n = 1, 2, ...,$ is uniformly distributed modulo m if and only if for all lattice points $\mathbf{h} \in \mathbb{Z}^k$ with $\mathbf{h} \neq 0 \pmod{m}$,*

$$\lim_{N\to\infty} \frac{1}{N} \sum_{n=1}^{N} e^{2\pi\sqrt{-1}\langle\mathbf{h},\mathbf{a}_n\rangle/m} = 0.$$

The notion of *k-distribution* of a sequence of integers, $a_n, n = 1, 2, ...,$ is also defined as follows:

DEFINITION **2.32** A sequence $a_n, n = 1, 2, ...,$ is said to be k-distributed modulo m if $\mathbf{a}_n = (a_n, ..., a_{n+k-1}), n = 1, 2, ...,$ is uniformly distributed modulo m. □

The above results and definitions are necessary in Chapter 3, where we describe the background of the spectral test for linear congruential sequences through Weyl's criterion.

Polynomial Case

Before introducing the uniform distribution for sequences of elements of $GF\{q,z\}$, we need the following definitions of *character*:

DEFINITION **2.33** Let α be in $GF(q)$, where $q = p^r$ for a prime p. Let $\omega_1, ..., \omega_r$ be a fixed basis for $GF(q)$ over $GF(p)$. Denote $\alpha = \sum_{i=1}^{r} a_i\omega_i$ with $a_1, ..., a_k \in GF(p)$. Then the character of α is defined by

$$e(\alpha; q) \stackrel{\text{def}}{=} e^{2\pi\sqrt{-1}a_1/p}.$$

□

DEFINITION **2.34** Let $\alpha(z) = c_w z^w + c_{w-1}z^{w-1} \cdots$ be in $GF\{q,z\}$, where w is an integer with $c_w \neq 0$ and $q = p^r$ for a prime p. Let $\omega_1, ..., \omega_r$ be a fixed basis for $GF(q)$ over $GF(p)$. Denote $c_{-1} = \sum_{i=1}^{r} a_i\omega_i$ with $a_1, ..., a_k \in GF(p)$. Then the character of $\alpha(z)$ is defined by

$$e(\alpha(z)) \stackrel{\text{def}}{=} e(c_{-1}; q) = e^{2\pi\sqrt{-1}a_1/p}.$$

□

For a sequence of elements of $GF\{q,z\}$, we have the following definition of uniform distribution, which corresponds to the uniform distribution modulo one of sequences of real numbers.

DEFINITION **2.35** Let $x_n(z), n = 1, 2, ...,$ be a sequence in $GF\{q,z\}$, and let $\alpha(z)$ be an arbitrary element in $GF\{q,z\}$. Let $\#(\alpha, r, N)$ denote the number of $x_n(z), n = 1, 2, ..., N$, such that the degree of the fractional part of $x_n(z) - \alpha(z)$ is less than $-r$. Then the sequence $x_n(z), n = 1, 2, ...,$ is said to be uniformly distributed modulo one in $GF\{q,z\}$ if

$$\lim_{N\to\infty} \frac{\#(\alpha, r, N)}{N} = q^{-r},$$

for all $r \geq 1$ and all $\alpha(z) \in GF\{q,z\}$ with $\nu(\alpha) \geq -r$. □

The Weyl criterion for a sequence in $GF\{q,z\}$ is given as follows:

Theorem 2.51 *A sequence $\alpha_n(z), n = 1, 2, \ldots$, is uniformly distributed modulo one in $GF\{q, z\}$ if and only if for all polynomials $h(z) \neq 0$,*

$$\lim_{N\to\infty} \frac{1}{N} \sum_{n=1}^{N} e(h(z)\alpha_n(z)) = 0.$$

Next, we overview the uniform distribution of sequences of polynomials over finite fields.

DEFINITION **2.36** Let $f_n(z), n = 1, 2, \ldots$, be a sequence of polynomials in $GF[q, z]$, and let $M(z)$ be a polynomial of degree r. For any polynomial $b(z)$ of degree less than r and any integer $N \geq 1$, let $\#(b(z), M(z), N)$ denote the number of elements among $f_1(z), \ldots, f_N(z)$ such that $f_n(z) \equiv b(z) \pmod{M(z)}$. Then the sequence $f_n(z), n = 1, 2, \ldots$, is said to be uniformly distributed modulo $M(z)$ if for all polynomials $b(z)$ of degree less than r,

$$\lim_{N\to\infty} \frac{\#(b(z), M(z), N)}{N} = q^{-r}.$$

□

Note that if a sequence of polynomials is uniformly distributed modulo $M(z)$, then it is also uniformly distributed modulo $m(z)$ for all $m(z)$ dividing $M(z)$. The Weyl criterion is as follows:

Theorem 2.52 *A sequence of polynomials, $f_n(z), n = 1, 2, \ldots$, over $GF(q)$ is uniformly distributed modulo $M(z)$ if and only if for all polynomials $h(z) \neq 0 \pmod{M(z)}$,*

$$\lim_{N\to\infty} \frac{1}{N} \sum_{n=1}^{N} e\left(\frac{h(z)f_n(z)}{M(z)}\right) = 0.$$

Let $a(z) = a_{r-1}z^{r-1} + \cdots + a_0$ and $b(z) = b_{r-1}z^{r-1} + \cdots + b_0$ be polynomials of degree less than r over $GF(q)$. Define $\langle a(z), b(z)\rangle = \sum_{i=0}^{r-1} a_i b_i$ in $GF(q)$; in other words, define it as the inner product of the coefficients. Then, the above theorem can be rewritten as follows:

Theorem 2.53 *A sequence of polynomials, $f_n(z), n = 1, 2, \ldots$, over $GF(q)$ is uniformly distributed modulo $M(z)$ if and only if for all polynomials $h(z) \neq 0$*

with $\deg(h) < \deg(M)$,

$$\lim_{N\to\infty} \frac{1}{N} \sum_{n=1}^{N} e(\langle \bar{f}_n(z), h(z) \rangle ; q) = 0,$$

where $\bar{f}_n(z) = f_n(z) \pmod{M(z)}$ *with* $\deg(\bar{f}_n) < \deg(M)$.

The special case of $q = 2$ becomes the most important in the next chapters. For this special case, the theorem is often called the XOR lemma in computer science literature (see Section 6.2).

The multidimensional version of the above results for the polynomial case is straightforward in the same way as the integer case.

2.3.2 Irregularities of Distribution

In this section, we will describe several important results in the theory of irregularities of distribution, which deal with the deviation of finite sequences from the ideal uniform distribution. We start with the definitions of *discrepancy.*

DEFINITION **2.37** Let $P = \{\mathbf{x}_n, n = 0, 1, ..., N-1\}$ be a set of points in $[0,1]^k$, and let $\mathbf{y} = (y_1, ..., y_k)$ be a point in $[0,1]^k$. $E(\mathbf{y})$ is defined as a subset of $[0,1]^k$, namely, $[0, y_1) \times \cdots \times [0, y_k)$. Let $\#(E(\mathbf{y}); N)$ denote the number of elements $\mathbf{x}_n, n = 0, 1, ..., N-1$, in $E(\mathbf{y})$. Then the discrepancy of the point set P is defined as

$$T_N^{(k)} = \left(\int_{[0,1]^k} \left(\frac{\#(E(\mathbf{y}); N)}{N} - \prod_{i=1}^{k} y_i \right)^2 d\mathbf{y} \right)^{1/2},$$

with respect to the L_2-norm, and

$$D_N^{(k)} = \sup_{\mathbf{y} \in [0,1]^k} \left| \frac{\#(E(\mathbf{y}); N)}{N} - \prod_{i=1}^{k} y_i \right|$$

with respect to the L_∞-norm. □

Obviously,

$$T_N^{(k)} \le D_N^{(k)}. \tag{2.7}$$

Note that

$$\lim_{N\to\infty} D_N^{(k)} = 0,$$

if and only if the sequence $\mathbf{x}_n, n = 0, 1, ...,$ is uniformly distributed in $[0,1]^k$. In one dimension ($k = 1$), we have explicit formulas for $T_N^{(1)}$ and $D_N^{(1)}$ [68, p.972], namely,

$$(T_N^{(1)})^2 = \frac{1}{12N^2} + \frac{1}{N}\sum_{n=0}^{N-1}\left(x_n - \frac{2n+1}{2N}\right)^2$$

and

$$D_N^{(1)} = \frac{1}{2N} + \max_{0\le n\le N-1}\left|x_n - \frac{2n+1}{2N}\right|.$$

Thus, we can show that for any $N \geq 1$, equally spaced points such as

$$x_n = \frac{n}{N} + \frac{1}{2N}, \quad \text{for } n = 0, 1, ..., N-1,$$

minimize both the L_2- and L_∞-discrepancies.

For general k dimensional case, i.e., $\mathbf{x}_n = (x_n^{(1)}, ..., x_n^{(k)}), n = 0, 1, ...,$ we have the following explicit formula for $T_N^{(k)}$:

$$(T_N^{(k)})^2 = \frac{1}{N^2}\sum_{n=0}^{N-1}\sum_{m=0}^{N-1}\prod_{i=1}^{k}(1-\max(x_n^{(i)}, x_m^{(i)})) - \frac{2^{1-k}}{N}\sum_{n=0}^{N-1}\prod_{i=1}^{k}(1-(x_n^{(i)})^2) + 3^{-k}. \tag{2.8}$$

From this, we have

Theorem 2.54 *The average of the square of $T_N^{(k)}$ over all sets of N points in $[0,1]^k$ is given by*

$$E((T_N^{(k)})^2) = \frac{2^{-k} - 3^{-k}}{N}.$$

The lower bound on $T_N^{(k)}$ is due to Roth [80].

Theorem 2.55 *For any integer $k \geq 1$, there exists a positive $c_1(k)$ such that for any integer $N > 1$ and for any set of N points in $[0,1]^k$, we have*

$$T_N^{(k)} > c_1(k)\frac{(\log N)^{(k-1)/2}}{N}.$$

Roth also showed that the above lower bound is sharp [81].

Theorem 2.56 *For any integer $k \geq 1$ and for a suitable $c_2(k)$, there exists, corresponding to any integer $N > 1$, a set of N points in $[0,1]^k$ such that*

$$T_N^{(k)} < c_2(k)\frac{(\log N)^{(k-1)/2}}{N}.$$

The proof of this theorem shows only the existence of such sequences; their explicit construction is still open.

For the case of L_∞-discrepancy $D_N^{(k)}$, the best lower bound currently known is obtained from that for the L_2-discrepancy $T_N^{(k)}$ by means of (2.7).

Theorem 2.57 *For any integer $k \geq 1$, there exists a positive $c_1(k)$ such that for any integer $N > 1$ and for any set of N points in $[0,1]^k$, we have*

$$D_N^{(k)} > c_1(k)\frac{(\log N)^{(k-1)/2}}{N}.$$

For $k = 2$ dimensions, W. M. Schmidt obtained a sharp lower bound (see [10]).

Theorem 2.58 *There exists a positive c such that for any integer $N > 1$ and for any set of N points in $[0,1]^2$, we have*

$$D_N^{(2)} > c\frac{\log N}{N}.$$

The problem of proving the sharp lower bound for the L_∞-discrepancy in $k > 2$ dimensions is called the "Great Open Problem," and may be stated as follows:

Conjecture 2.1 *For any integer $k \geq 1$, there exists a positive $c_3(k)$ such that for any integer $N > 1$ and for any set of N points in $[0,1]^k$, we have*

$$D_N^{(k)} > c_3(k)\frac{(\log N)^{k-1}}{N}.$$

For the average behavior of $D_N^{(k)}$, the law of iterated logarithms shows (see Niederreiter [72]) that

Theorem 2.59 *For a set of uniformly random N points in $[0,1]^k$, we have*

$$\limsup_{N\to\infty} \frac{\sqrt{2N}D_N^{(k)}}{\sqrt{\log\log N}} = 1$$

with probability one.

As will be shown in Chapter 6, *low-discrepancy sequences* play the most important roles in quasi-Monte Carlo methods for numerical integration and global optimization. We therefore give definitions of them.

DEFINITION **2.38** We define a *low-discrepancy sequence* in dimension k as an (infinite) sequence $\mathbf{x}_0, \mathbf{x}_1, ...,$ in $[0,1]^k$ such that for all $N > 1$, the discrepancy of the first N points is

$$D_N^{(k)} \le c(k)\frac{(\log N)^k}{N},$$

where $c(k)$ is a constant depending only on the dimension k. □

Once any low-discrepancy sequences are given, we can construct point sets whose discrepancy bound attains the same order as the conjectured lower bound of L_∞-discrepancy.

Theorem 2.60 *Let $\mathbf{x}_n = (x_{n,1}, ..., x_{n,k-1}), n = 0,1,...,$ be a low-discrepancy sequence in $k-1$ dimensions. Then for any integer $N > 1$ the set of N points, $\mathbf{x}_n = (x_{n,1}, ..., x_{n,k-1}, n/N), n = 0,1,..., N-1$, satisfies*

$$D_N^{(k)} \le c(k)\frac{(\log N)^{k-1}}{N},$$

where $c(k)$ is a constant depending only on the dimension k.

A classical family of low-discrepancy sequences are Halton sequences [32], which are based on the radical inverse function defined as follows:

DEFINITION **2.39** For integers $n \ge 0$ and $b \ge 2$,

$$\phi_b(n) = \frac{a_0}{b} + \frac{a_1}{b^2} + ... + \frac{a_m}{b^{m+1}},$$

where the digital expansion of n in base b is given as $a_m b^m + \cdots + a_1 b + a_0$ with integers a_j, $0 \le a_j < b$ for $0 \le j \le m$, and $m = [\log_b n]$. □

DEFINITION **2.40** The sequence, $\phi_2(n), n = 0, 1, \ldots$, is called the van der Corput sequence. □

The k-dimensional generalization of the van der Corput sequence is called Halton sequences.

DEFINITION **2.41** The Halton sequence $\mathbf{x}_n, n = 0, 1, \ldots$, in $[0, 1]^k$ is defined as

$$\mathbf{x}_n = (\phi_{b_1}(n), \ldots, \phi_{b_k}(n)),$$

where $b_1, \ldots, b_k$ are pairwise coprime. □

The discrepancy of the Halton sequence is given below.

Theorem 2.61 *For any integer $N > 1$, the discrepancy of the first N points of the Halton sequence is given as*

$$D_N^{(k)} \leq c(b_1, \ldots, b_k)\frac{(\log N)^k}{N} + O\left(\frac{(\log N)^{k-1}}{N}\right),$$

where $c(b_1, \ldots, b_k) \approx \Pi_{i=1}^k \frac{b_i}{\log b_i}$.

Applying Theorem 2.60 to the Halton sequence, we obtain the so-called Hammersley N-point set,

$$\mathbf{x}_n = (\phi_{b_1}(n), \ldots, \phi_{b_{k-1}}(n), \frac{n}{N})$$

for $n = 0, 1, \ldots, N-1$, where $b_1, \ldots, b_{k-1}$ are pairwise coprime.

Generalized Halton sequences, which sometimes is called scrambled Halton sequences, are important in practical applications.

DEFINITION **2.42** For $n \geq 0$, define

$$\phi_b(n; \pi) \equiv \frac{\pi_0(a_0)}{b} + \frac{\pi_1(a_1)}{b^2} + \ldots + \frac{\pi_m(a_m)}{b^{m+1}},$$

where $\pi = (\pi_0, \pi_1, \ldots)$ is a set of permutations of $\{0, 1, \ldots, b-1\}$, and $a_j, j = 0, 1, \ldots, m$, are coefficients of the digital expansion of n in base b, that is, $n =$

$a_0 + a_1 b + \cdots + a_m b^m$ with $m = [\log_b n]$. Then, a generalized Halton sequence $\mathbf{x}_n, n = 0, 1, ...,$ in $[0,1]^k$ is defined as

$$\mathbf{x}_n = (\phi_{b_1}(n; \pi^{(1)}), ..., \phi_{b_k}(n; \pi^{(k)})),$$

where $b_1, ..., b_k$ are pairwise coprime. □

The original Halton sequence corresponds to the case where all the permutations are the identity mapping. Warnock [106] used $\pi_m^{(i)}(a) = a + m \pmod{b_i}$, $m = 0, 1, 2, ...,$ for $1 \le i \le k$.

Finally, we give a general upper bound for the discrepancy of any point set with rational coordinates in the k-dimensional unit cube. This result gives an important relation between discrepancy and Weyl's criterion; that is, it shows the bound in terms of the exponential sum, which appears in Weyl's criterion. Define

$$S_{max}(N) = \max_{\substack{\mathbf{h} \in C_k(M) \\ \mathbf{h} \neq \mathbf{0}}} \left| \frac{1}{N} \sum_{n=0}^{N-1} e^{2\pi\sqrt{-1}\langle \mathbf{h}, \mathbf{x}_n \rangle / M} \right|,$$

where $C_k(M) = \{(i_1, ..., i_k) \in \mathbb{Z}^k \mid -M/2 < i_1, ..., i_k \le M/2\}$. Then we have the following theorem [72, Corollary 3.11]:

Theorem 2.62 *For integers $M > 1$ and $N > 1$, define the point set $P = \{\{\mathbf{x}_n / M\}, n = 0, 1, ..., N-1 \mid \mathbf{x}_n \in \mathbb{Z}^k\}$, where $\{(x_1, ..., x_k)\}$ denotes the fractional part of all coordinates, $(\{x_1\}, ..., \{x_k\})$. Then we have*

$$1 - \left(1 - \frac{1}{M}\right)^k \le D_N^{(k)} \le 1 - \left(1 - \frac{1}{M}\right)^k + c(k) S_{max}(N) (\log M)^k,$$

where $c(k)$ is a constant depending only on k.

The term $1 - \left(1 - \frac{1}{M}\right)^k$ is called the discretization error, which is a general (trivial) lower bound for the discrepancy of point sets consisting entirely of points with rational coordinates whose denominators are at most M.

* * *

Notes and Remarks

For an introduction to elementary number theory and finite fields relevant to the materials in Section 2.1, the reader is referred to books by Hardy and Wright [33] and Lidl and Niederreiter [51].

On the geometry of numbers, the reader should consult Gruber and Lekkerkerker [31] and Pohst and Zassenhaus [76] for more details and proofs of many important theorems. More sophisticated algorithms than SOLVEQX and LLL-BASIS are found in Cohen [20].

The most comprehensive monograph on uniform distribution of sequences is by Kuipers and Niederreiter [44]. The relation between the notion of ∞-distributivity and randomness is fully discussed in Knuth [40]. Recent results on irregularities of distribution are found in the two excellent books by Beck and Chen [10] and Niederreiter [72].

3

LINEAR CONGRUENTIAL GENERATORS

Random numbers fall mainly in the planes.
– George Marsaglia (1968)

The linear congruential method is the most popular algorithm for random number generation in the field of computer simulations. The linear congruential generator (LCG) is defined by

$$\begin{aligned} X_n &= aX_{n-1} + c \pmod{M}, \\ u_n &= \frac{X_n}{M}, \end{aligned}$$

where a, c, and M are integers and $X_n, n = 1, 2, ...,$ is a sequence of integers with $0 \le X_n < M$. The normalized sequence $u_n, n = 1, 2, ...,$ is a random number sequence in $[0, 1)$. The following results tell us how to choose the parameters a, c, and M to obtain the maximal period lengths.

For the case $c \neq 0$, we have

Theorem 3.1 *An LCG with the parameters (a, c, M) has period length M if and only if*

(i) $\gcd(c, M) = 1$;

(ii) $a = 1 \pmod{p}$ *for every prime p dividing M;*

(iii) $a = 1 \pmod{4}$ *if M is a multiple of 4.*

For the case $c = 0$, which is more important in practice, we have

Theorem 3.2 *An LCG with the parameters $(a, 0, M)$ has a maximal period length $\lambda(M)$ if and only if*

(i) $\gcd(X_0, M) = 1$;

(ii) *a is a primitive element modulo M.*

In the 1960's, the most popular choice of parameters was $c = 0$ and $M = 2^w, w \geq 4$; in this case $a = 3$ or $5 \pmod 8$ is the necessary and sufficient condition for the maximal period. (For example, RANDU, IBM scientific library subroutine widely used all over the world in the 1960s, used the parameters $a = 65539, c = 0$, and $M = 2^{31}$.) However, it is now known that this choice is problematic because of the non-randomness of the low-order bits of X_n. To be precise, the least significant l bits of $X_n, n = 1, 2, ...$, have a period of at most 2^l. Recently, therefore, a common choice has been to take M as prime and a as a primitive root modulo M. One of the most popular generators of this type is the so-called GGL, the IBM random number subroutine [49], in which $a = 7^5$, $c = 0$, and $M = 2^{31} - 1$. We need to notice that if M is a prime and a is primitive modulo M, then the period becomes $M - 1$ whatever c is. This is because if we put $Y_n = X_n + c/(a-1)$, we have a homogeneous form $Y_n = aY_{n-1} \pmod M$. Notice also that for a prime M the maximum period M is attained only when $a = 1$ and $\gcd(c, M) = 1$, which, unfortunately, produces a non-random-looking sequence, such as $1, 2, 3, ...$, when $c = 1$.

On an L-bit word computer (L is usually 32), the multiplication of X by a often causes an overflow, since all single-precision integers are between -2^{L-1} and $2^{L-1}-1$. We therefore need a technique to avoid any overflow in the calculation of $aX \pmod M$. One idea is based on the following: Let $q = [M/a]$ and $r = |M|_a$. Then we have

$$aX = a|X|_q - \left[\frac{X}{q}\right] r \pmod M.$$

If $ar < M$, it is easy to see that no overflow occur when we use the RHS. On the basis of this idea, Schrage (see [16]) implemented the generator GGL in an efficient and portable way.

In many applications, we need multiple independent streams of random number sequences. One approach to handling this problem is to partition an entire period of a single sequence into many long disjoint subsequences. In order to do that, a random number generator must quickly jump ahead to the beginnings of different subsequences and jump back to the previous sequence. This property is called the 'jumping ahead' property, and is easily implemented for LCGs as follows: Assume that we need m disjoint subsequences from an LCG. Let $s_1, ..., s_m$ be the seed values for the subsequences. Then we can set the seed values as follows: for $i = 1, 2, ..., m-1$

$$s_{i+1} = a^j s_i \pmod{M},$$

where j should be chosen very large, to avoid overlapping, and should also be such that $a^j \pmod{M}$ is a 'good' multiplier in order to avoid bad correlation properties between the disjoint subsequences.

Since the word-length of almost all computers is 32 bits, the modulus M of the LCG is usually chosen less than 2^{32}. Thus, the period length of the resulting random number sequence is also at most 2^{32}, which can now be exhausted in a few minutes of CPU time even on personal computers. The most popular way to obtain much longer periods of sequences is the multiple recursive generator (MRG), an extended version of the LCG, which is defined by, for $n = r+1, r+2, ...,$

$$X_n = a_{r-1}X_{n-1} + \cdots + a_0 X_{n-r} \pmod{M},$$

where the initial values $X_1, ..., X_r$ are not all-zero, and the coefficients $a_0, ..., a_{r-1}$ are chosen so that the period length of $X_n, n = 1, 2, ...,$ becomes very large. If $a_i, i = 0, ..., r-1$, are all chosen so as to satisfy the condition $a_i|M|_i < M$, the multiplication overflow can be avoided in software implementation. There are two popular choices of M: a prime or 2^w, typically $w \approx 32$. If M is a prime, the characteristic polynomial of $X_n, n = 1, 2, ...,$ is usually chosen as a primitive trinomial of the form $z^r - a_{r-s}z^s - a_0 (r > s)$ over $GF(M)$, and thus the sequence is produced by using the trinomial-type recurrence relation

$$X_n = a_{r-s}X_{n-s} + a_0 X_{n-r} \pmod{M}.$$

This scheme has two advantages in the light of implementation: (1) the resulting period length becomes the maximum, $M^r - 1$, and (2) the generation speed is fast, since one random number can be generated by two modulo-multiplications and one modulo-addition.

If $M = 2^w$, the so-called additive and subtractive generators, which some people call lagged Fibonacci generators with + and −, have been recommended by Knuth [40]: for $n = r, r+1 ...,$ the sequence is produced by

$$X_n = X_{n-s} \pm X_{n-r} \pmod{2^w}. \tag{3.1}$$

Unfortunately, in this case, the maximum period length is $2^{w-1}(2^r - 1)$, which is very short compared with 2^{rw}, the maximum possible period using the same size of memory area, although the generation speed is very fast. The following theorem due to Brent [17] is very helpful for choosing the parameters r, s, and w.

Theorem 3.3 *If $z^r - a_{r-s}z^s - a_0$ $(r > s)$ is a primitive trinomial over $GF(2)$ with $r > 2$ and if $X_0, ..., X_{r-1}$ are not all even, then the sequence $X_n, n = 0, 1, ...,$ defined in (3.1) has the maximum period $2^{w-1}(2^r - 1)$.*

Table 4.1 contains primitive trinomials over $GF(2)$ with Mersenne prime periods.

In this chapter, we start with a description of *lattice structure* underlying the sequences generated by LCGs as well as by MRGs, and then explain in detail the *spectral test*, which has been widely used as a powerful theoretical test for examining the uniformity of lattice structures associated with LCGs and MRGs. From this viewpoint, we show that sequences generated by trinomial-type MRGs described in the above suffer from 'bad' lattice structures in high dimensions. Finally, we present two schemes that efficiently produce very long periods of linear congruential sequences: Marsaglia-Zaman and Wichmann-Hill schemes, and then discuss how to construct long-period LCGs with 'good' lattice structures.

3.1 LATTICE STRUCTURE

Consider the k-tuples $(X_{j_1+n}/M, X_{j_2+n}/M, ..., X_{j_k+n}/M), n = 1, 2, ...,$ from an LCG, where $j_1, ..., j_k$ are integers with $0 \leq j_1 < \cdots < j_k \leq M - 1$. It is then easily shown that all these k-tuples are contained in $[0, 1)^k \cap (\mathcal{L}_k + \lambda)$, where $\mathcal{L}_k$ is the lattice whose basis are

$$\begin{aligned} \mathbf{e}_1 &= \frac{1}{M}(1, a^{j_2-j_1}, \ldots, a^{j_k-j_1}), \\ \mathbf{e}_2 &= (0, 1, 0, \ldots, 0), \\ &\vdots \\ \mathbf{e}_k &= (0, 0, 0, \ldots, 1), \end{aligned} \tag{3.2}$$

and $\lambda = \frac{c}{M(a-1)}(0, a^{j_2-j_1} - 1, ..., a^{j_k-j_1} - 1)$. This is a well-known lattice structure of linear congruential sequences. Note that the lattice $\mathcal{L}_k$ contains $\mathbb{Z}^k$; in other words, the lattice $\mathcal{L}_k$ is cyclic modulo one. Thus, all structural information about the lattice is contained in $[0, 1)^k$. Clearly, this fact is common to all linear congruential sequences in every dimension. Although, in most studies, only the case $(j_1, ..., j_k) = (1, ..., k)$ has been considered, here we should emphasize that for any k-tuple $(j_1, ..., j_k)$, linear congruential sequences have (integer) lattice structures.

Figure 3.1 shows the two-dimensional plot of (u_n, u_{n+1}) from the linear congruential sequence $u_n = X_n/509$ with

$$X_n = 170X_{n-1} \pmod{509} \text{ for } n = 1, 2, \ldots.$$

We see that all the points lie on only three lines. There are many sets of parallel lines that can cover all the points, but in this case, three is the minimum number of lines in such sets. Keeping the modulus M the same, that is, 509, and varying the multiplier $a, 0 < a < 509$, so that it is primitive modulo M, we can investigate how large the minimum number of lines covering all the points can be made over all the choices of multipliers. It turns out that the minimum number can be at most 30. The complete set of multipliers corresponding to the maximum of 30 lines is given as $a = \pm 35, \pm 59, \pm 69, +89, \pm 160$, and -183 modulo 509. Figure 3.2 shows the two-dimensional plot for $a = 59$, which is much more uniform than Figure 3.1.

For the case of an rth order MRG sequence $X_n, n = 1, 2, ...,$ with the maximum period $M^r - 1$, it is obvious that the point set $\{(X_{j_1+n}, X_{j_2+n}, ..., X_{j_k+n}), n =$

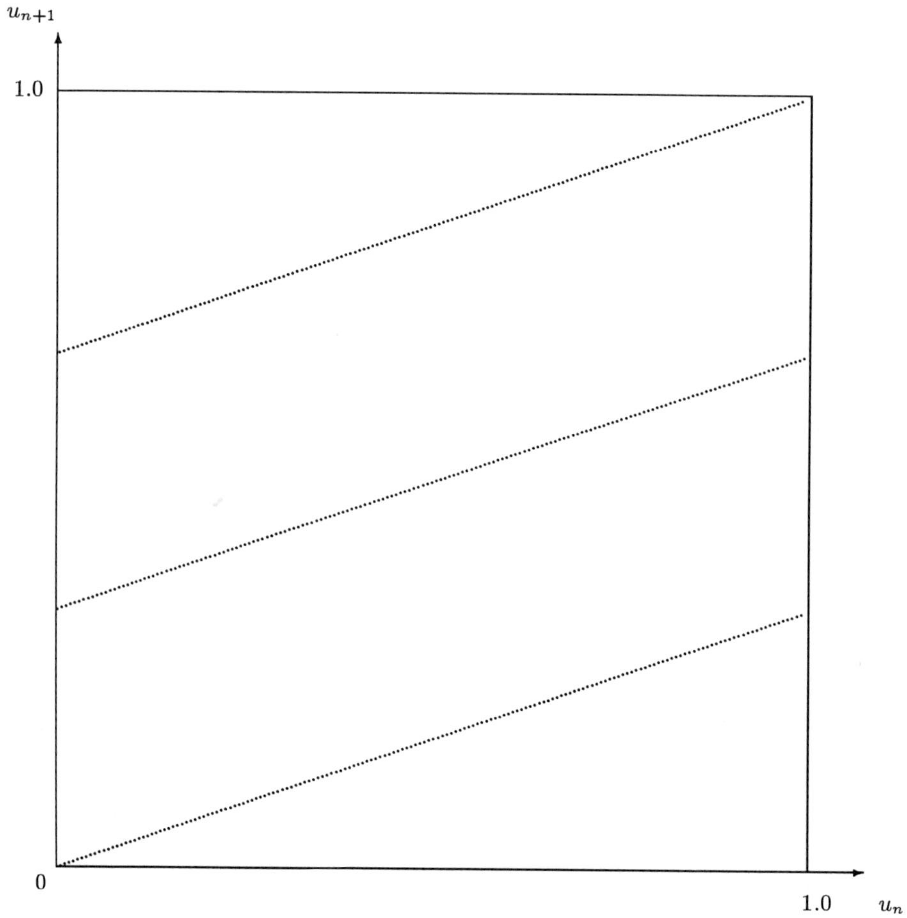

Figure 3.1 A set of two-dimensional points, $(u_n, u_{n+1}), n = 1, ..., 508$, produced by $u_n = X_n/509$, where the LCG is given as $X_n = 170X_{n-1} \pmod{509}$.

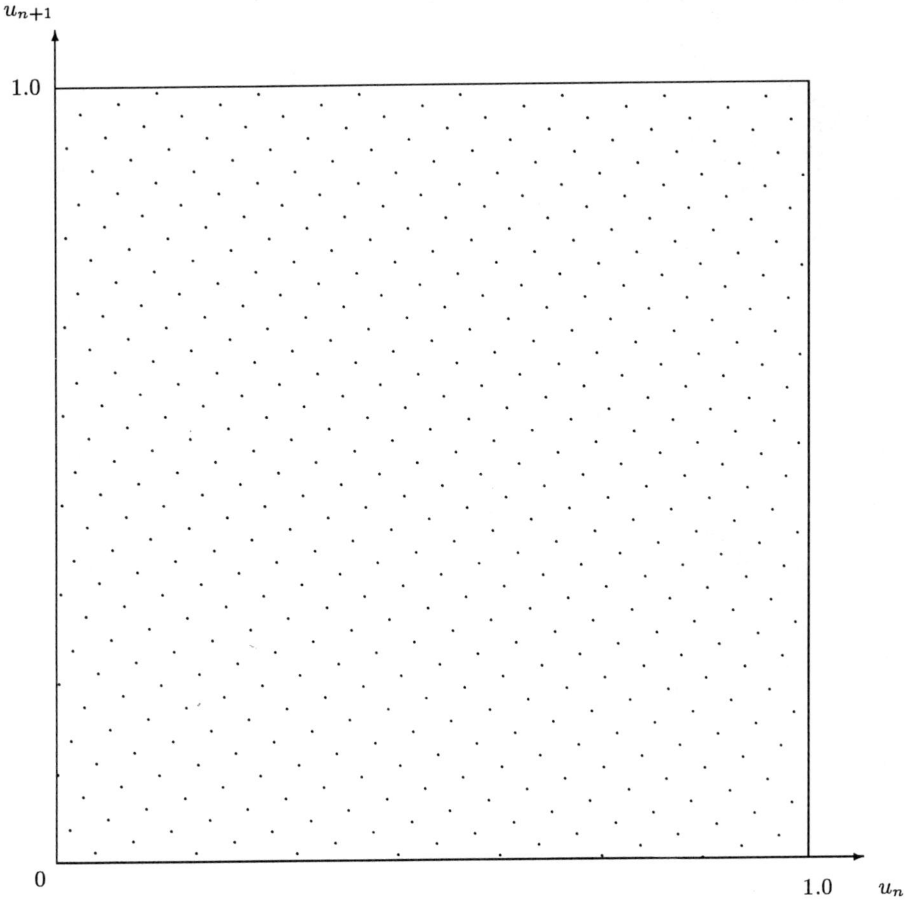

Figure 3.2 A set of two-dimensional points, $(u_n, u_{n+1}), n = 1, ..., 508$, produced by $u_n = X_n/509$, where the LCG is given as $X_n = 59X_{n-1} \pmod{509}$.

$1, 2, \ldots\}$ is almost equidistributed[1] for $k \le r$ if M is prime and $j_1, \ldots, j_k$ are chosen so that $\mathbf{x}_0 C^{j_i}, i = 1, 2, \ldots, k$, are linearly independent over $GF(M)$, where C is the $r \times r$ companion matrix of the recurrence relation for the MRG and $\mathbf{x}_0 = (X_1, X_2, \ldots, X_r)$ is an arbitrary nonzero vector over $GF(M)$. Note that the matrix C is nonsingular. For the case $r < k$, the lattice structure of MRG is described as follows: Let $\mathcal{A}$ be an $r \times k$ matrix over $GF(M)$ such that

$$(X_{j_1+n}, X_{j_2+n}, \ldots, X_{j_k+n}) = (X_1, X_2, \ldots, X_r)C^n \mathcal{A}.$$

Consider the case $k = dr$, and denote $\mathcal{A} = (A_1 A_2 \cdots A_d)$, where $A_i, i = 1, \ldots, d$, are $r \times r$ matrices. If $j_1, \ldots, j_k$ are chosen so that $A_i, i = 1, \ldots, d$, are nonsingular, then in $k = dr$ dimensions the point set has the so-called lattice structure whose bases are given as the row vectors of the $dr \times dr$ matrix

$$\begin{pmatrix} I & G_1^{-1}G_2 & G_1^{-1}G_3 & \cdots & G_1^{-1}G_d \\ 0 & MI & 0 & 0 & 0 \\ 0 & 0 & MI & 0 & 0 \\ \vdots & & \vdots & & \vdots \\ 0 & 0 & 0 & 0 & MI \end{pmatrix}, \tag{3.3}$$

where $G_i = A_i^{-1} C A_i, 1 \le i \le d$, and I is the $r \times r$ identity matrix. When we consider the k-dimensional case with $dr < k < (d+1)r$, the last $k - dr$ rows and the last $k - dr$ columns of the matrix should be omitted [1].

A quick way to check the lattice structure of linear congruential sequences is based on *potency* [40], which is defined as the smallest (positive) integer s satisfying $(a-1)^s = 0 \pmod{M}$ if it exists; otherwise ∞. For the maximum period LCG, it follows from Theorem 3.1 that the potency is 1 if M is square free, and that it is at most $\lceil e/2 \rceil$ if $M = 2^e$ for $e > 1$. The notion of potency is interpreted as follows: For a linear congruential sequence, we have the relation

$$X_{n+1} - X_n = a^n c \pmod{M},$$

where $c = X_1 - X_0$. By putting $d = a - 1$, we get

$$X_{n+1} - X_n = \sum_{j=0}^{s-1} \binom{n}{j} d^j c \pmod{M}.$$

For example, consider $s = 3$. By differencing both sides of the equation twice, we obtain

$$X_{n+1} - 3X_n + 3X_{n-1} - X_{n-2} = cd^2 \pmod{M}.$$

[1] Here, the almost equidistribution of k-dimensional integer vectors, $(X_n^{(1)}, \ldots, X_n^{(k)}), n = 1, 2, \ldots, N$, with $N \approx M^r$, means that every k-tuple of integers between 0 and $M-1$ appears approximately $N/M^k \approx M^{r-k}$ times, where $0 \le X_n^{(j)} \le M - 1$ for $j = 1, \ldots, k(\le r)$.

This means that all points, $(X_{n+1}, X_n, X_{n-1}, X_{n-2}), n = 1, 2, ...,$ lie on hyperplanes defined by this equation. In general, if a set of parallel hyperplanes in the k-dimensional Euclidean space is expressed by the formula

$$h_1 x_1 + \cdots + h_k x_k = h_0 \quad (\text{mod} M),$$

where $h_0, h_1, ..., h_k$ are fixed integers, then the number of hyperplanes in $[0, M-1]^k$ is no more than

$$|h_1| + \cdots + |h_k|.$$

Hence, in the above case of $s = 3$, the number of hyperplanes is at most $1+3+3+1 = 8$. This value is significantly small. Thus we can conclude that LCGs with low potency should not be used.

3.2 SPECTRAL TEST

The lattice structure of linear congruential sequences can be analyzed by means of the so-called spectral test [23], which provides a measure defined as the length of the shortest vector in the dual of the lattice spanned by the linear congruential sequence. This measure has proved very powerful in examining the resolution of uniformity of linear congruential sequences since Knuth's comprehensive study of the test [40]. So far, most cases to have been considered in analysis of the lattice structure of a particular sequence are on that of the k-tuple, $(j_1, ..., j_k) = (1, ..., k)$, namely, the consecutive k elements of linear congruential sequences. However, we emphasize that the spectral test can be applied to any k-tuples. Hereafter, without loss of generality, we assume that $j_1 = 0 < j_2 < \cdots < j_k < M$.

The measure provided by the spectral test is defined as follows: for k dimensions,

$$\nu_k = \min \left(\sum_{i=1}^{k} s_i^2 \right)^{1/2}, \tag{3.4}$$

where the minimum is taken over all nonzero solutions of the equation

$$\sum_{i=1}^{k} a^{j_i} s_i = 0 \quad (\text{mod} M). \tag{3.5}$$

An intuitive interpretation of this measure is that ν_k^{-1} is the maximum distance between hyperplanes, taken over all families of hyperplanes covering all

k-dimensional points generated by the k-tuples from a linear congruential sequence. Therefore, $\log_2 \nu_k$ gives the resolution of uniformity for a given sequence. In other words, ν_k can be regarded as the length of the nonzero shortest vector in the lattice spanned by the following vectors:

$$\begin{aligned} \mathbf{e}_1^* &= (M, 0, 0, ..., 0), \\ \mathbf{e}_2^* &= (-a^{j_2 - j_1}, 1, 0, ..., 0), \\ &\vdots \\ \mathbf{e}_k^* &= (-a^{j_k - j_1}, 0, 0, ..., 1). \end{aligned}$$

Note that this lattice is the dual of the lattice given by (3.1). In brief, for a given set of a, c, and M, the larger the value of ν_k, the more uniform the sequence is. Note that the value of c has no effect on the result of the spectral test, although the period length of the resulting sequence is influenced by the choice of c.

The minimum number of hyperplanes covering all k-dimensional points from a linear congruential sequence has also been considered as a measure for the uniform distribution of linear congruential sequences. As Marsaglia explained [55], this measure can be formalized as the length of the shortest nonzero vector w.r.t. the L_1-norm in the dual lattice, namely

$$\tilde{\nu}_k = \min \sum_{i=1}^{k} |s_i|, \tag{3.6}$$

whereas the definition in (3.4) is based on the L_2-norm. Therefore, the only difference is the choice of the norm. For the lattice structure of the LCG in Figures 3.2, the value of ν_2^{-1} is $\frac{1}{\sqrt{514}} = .0441..$, given by the equation $-15u_i - 17u_{i+1} = 0 \pmod 1$, whereas the minimum number of hyperplanes covering all the points is 30, given by the equation $-22u_i + 9u_{i+1} = 0 \pmod 1$.

We recall a famous story about RANDU, which is formulated as

$$X_n = 65539 X_{n-1} \pmod{2^{31}},$$

where we should notice that $65539 = 2^{16} + 2 + 1$. By an easy hand calculation, we have the following equation of hyperplanes:

$$9X_n - 6X_{n+1} + X_{n+2} = 0 \pmod{2^{31}}.$$

Thus, the value of ν_k in three dimensions is

$$\nu_3 = \sqrt{9^2 + (-6)^2 + 1^2} = \sqrt{118}.$$

Since we can easily check that this is the maximum distance over all families of hyperplanes, we have $\nu_k^{-1} = \frac{1}{\sqrt{118}} = 0.092...$, and we also obtain that the minimum number of hyperplanes is $(9+6+1)-1 = 15$. This means that the uniform distribution in three dimensions of the sequence from RANDU is very bad.

As shown in Section 2.2, there are several important results about the successive minima of primal and dual lattices. Note that ν_k is equal to the first successive minimum of the dual lattice associated with linear congruential sequences. Ripley [77] considered the ratio of the first and last successive minima, denoted by l_1 and l_k, as a measure for the uniformity of lattice structures of linear congruential sequences. Intuitively speaking, the ratio l_1/l_k can be viewed as a measure for the granularity of a lattice (see Figure 3.3). If the ratio is close to 1, the basic parallelogram spanning the lattice looks like a k-dimensional 'cubic', and thus the lattice points are distributed almost uniformly. He obtained the following results:

Theorem 3.4 *For $k > 1$, we have*

$$1 \le \nu_k l_k \le \gamma_k^k,$$

where γ_k is the Hermite's constant (see Section 2.2).

This means that l_k is very close to ν_k^{-1} in small dimensions. The next result gives a relation between the spectral test and Ripley's ratio.

Theorem 3.5 *We have*

$$\frac{\nu_k^k}{\gamma_k^k M} \le \frac{l_1}{l_k} \le 1,$$

for $k \ge 2$, where M is the modulus of an LCG.

The so-called *Beyer's ratio* employs a slightly different measure, which is defined as the ratio of the length of the shortest reduced base to that of the longest one. Its intuitive interpretation is the same as that of Ripley's ratio. In fact, for $k \le 4$, the ratios are equivalent due to Theorem 2.34. For $k > 4$ dimensions, Beyer's ratio is, in general, a bit smaller than Ripley's. As explained in Section 2.2, the computation of successive minima will be a bit easier than that of reduced bases.

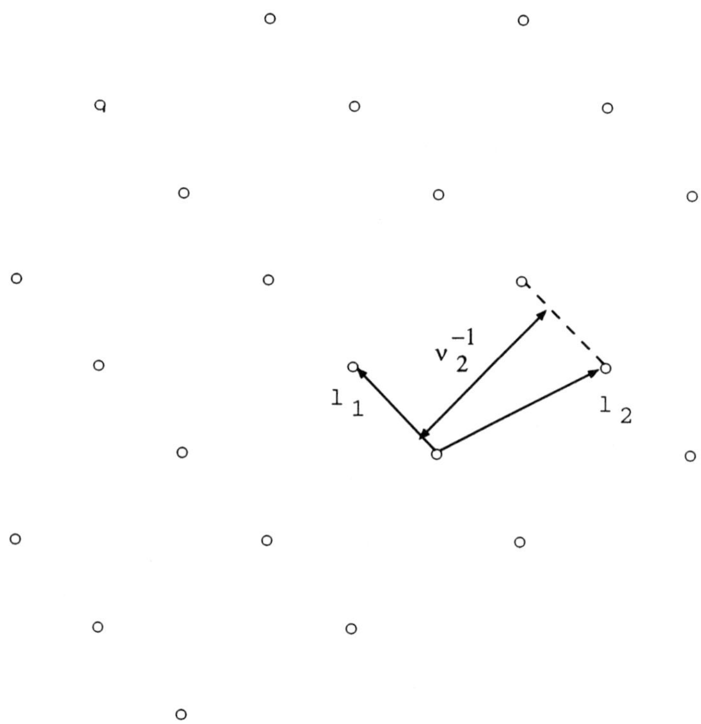

Figure 3.3 Illustration of the difference between the spectral test and Ripley's ratio.

There is an important connection between the finite, discrete version of Weyl's criterion and the spectral test. As we have already explained, the spectral test provides the length of the shortest vector in the dual of the lattice associated with a linear congruential sequence. In terms of the exponential sum, this is written as follows:

$$\nu_k = \min_{\substack{\mathbf{h}\in C_k(M)\\ \mathbf{h}\neq 0}} \left\{ |\mathbf{h}| \;\middle|\; \left| \frac{1}{N}\sum_{n=1}^{N} e^{2\pi\sqrt{-1}\langle \mathbf{h},\mathbf{x}_n\rangle/M} \right| = 1 \right\},$$

where $C_k(M) = \{(h_1, ..., h_k) \in \mathbb{Z}^k \mid -M/2 < h_1, ..., h_k \leq M/2\}$ and $\mathbf{x}_n = (X_{j_1+n}, ..., X_{j_k+n}), n = 1, 2, ...$ Note that the condition

$$\left| \frac{1}{N}\sum_{n=1}^{N} e^{2\pi\sqrt{-1}\langle \mathbf{h},\mathbf{x}_n\rangle/M} \right| = 1$$

is equivalent to that $\langle \mathbf{h}, \mathbf{x}_n\rangle$ is constant modulo M for all $n = 1, ..., N$, which directly corresponds to the condition (3.5). In Chapter 3 of the first edition of his book [40], Knuth gave an interesting interpretation of the spectral test with respect to Weyl's criterion. According to this interpretation, if the value of $|\frac{1}{N}\sum_{n=1}^{N} e^{2\pi\sqrt{-1}\langle \mathbf{h},\mathbf{x}_n\rangle/M}|$ is close to 1 for some nonzero $\mathbf{h}'$, there exist some 'non-random' (or non-uniform) properties in the sequence $\mathbf{x}_n, n = 1, ..., N$, from the viewpoint of Fourier analysis. He therefore claimed that such $|\mathbf{h}'|$ should be as large as possible so as to avoid non-random properties. This exactly coincides with the intuitive interpretation of the spectral test that ν_k^{-1}, the maximum distance between hyperplanes, should be as small as possible.

Finally, we should mention that some popular MRG sequences suffer from inherent bad lattice structure due to the simplicity of their generation scheme. First, consider additive and subtractive generators. From the definition of the generators, all the points $(X_n, ..., X_{n+r}), n = 1, 2, ...,$ lie on the hyperplanes in $r + 1$ dimensions given by

$$X_{n+r} - X_{n+s} \pm X_n = 0 \pmod{M}.$$

It follows immediately from this equation that the maximum distance between hyperplanes becomes at least $1/\sqrt{3}$. This means that the lattice structure of this class of generators in higher dimensions than r is very bad. In other words, the triple $(X_n, X_{n+s}, X_{n+r}), n = 1, 2, ...,$ from the generators has a bad lattice structure in three dimensions; that is, the third-order correlation with $(j_1, j_2, j_3) = (0, s, r)$ of the sequence is far from that of (truly) random sequences. In fact, Coddington [19] reported that these generators produce unsatisfactory biased results when used in high-precision Ising model simulations. Another type of generator that we should be careful to use is a class of

MRGs with primitive trinomials, which have been recommended by some people mainly because they can be efficiently implemented. However, it is known that it is difficult [45] to find such 'simplistic' generators with good lattice structures in terms of the spectral test as well as Beyer's ratio. There must be some trade-off between the implementation cost and the good uniform distribution.

3.3 IMPLEMENTING LONG-PERIOD LCGS

In this section, we describe how to efficiently implement long-period linear congruential generators. There are two ways. The first is to use large prime moduli of special form. The idea of this approach comes from *add-with-carry* (AWC) and *subtract-with-borrow* (SWB) generators, which were proposed by Marsaglia and Zaman [59]. The second way is to combine several LCGs. This idea stems from Wichmann-Hill combined generators [108], which have also drawn considerable attention from practical users [46] in the past decade because of their efficiency and portability.

We should mention an important remark about the lattice structure of long-period LCGs. In practice, a random number u_n in $[0, 1)$ is obtained through normalization by M as a fractional expansion with base b:

$$u_n = \frac{X_n}{M} = \frac{x_1^{(n)}}{b} + \frac{x_2^{(n)}}{b^2} + \cdots + \frac{x_L^{(n)}}{b^L} + \cdots,$$

where usually $b = 2$ on binary computers. In such a situation, what we actually use is, precisely speaking, a truncated version of u_n. Let $\tilde{u}_n$ be the truncation of u_n; that is, for some L,

$$\tilde{u}_n = \frac{x_1^{(n)}}{b} + \frac{x_2^{(n)}}{b^2} + \cdots + \frac{x_L^{(n)}}{b^L}.$$

If we form the k-dimensional vectors as

$$P_{k,n} = (u_n, ..., u_{n+k-1}),$$

and

$$\tilde{P}_{k,n} = (\tilde{u}_n, ..., \tilde{u}_{n+k-1}),$$

then the Euclidean distance between $P_{k,n}$ and $\tilde{P}_{k,n}$ is bounded by $\sqrt{k}/b^L$, which is typically around 2^{-30} for a 32-bit computer. We know that the points $P_{k,n}, n = 1, 2, ...,$ form a lattice structure and that in high dimensions $\sqrt{k}/b^L$ could be small relative to the distance between hyperplanes. Thus, the hyperplane structure of $P_{k,n}, n = 1, 2, ...,$ could give a good approximation of the behavior of the points $\tilde{P}_{k,n}, n = 1, 2, ...,$ which we actually use on a computer for our applications.

3.3.1 LCGs with Large Prime Moduli

We start with the definition of *add-with-carry* (AWC) and *subtract-with-borrow* (SWB) generators: Let I denote the indicator function, whose value is 1 if its argument is true, and 0 otherwise. Assume that $r > s$. The AWC is defined by

$$\begin{aligned} x_n &= (x_{n-s} + x_{n-r} + c_n) \bmod b, & (3.7)\\ c_{n+1} &= I(x_{n-s} + x_{n-r} + c_n \geq b). & (3.8) \end{aligned}$$

A variant of AWC, called *complementary* AWC, or AWC-c, is given by:

$$\begin{aligned} x_n &= (2b - 1 - x_{n-s} - x_{n-r} - c_n) \bmod b & (3.9)\\ &= (-x_{n-s} - x_{n-r} - c_n - 1) \bmod b, & \\ c_{n+1} &= I(x_{n-s} + x_{n-r} + c_n \geq b). & (3.10) \end{aligned}$$

Here, c_n is called the *carry*. The SWB also comes in two flavors, which we will call SWB-I and SWB-II, based on the recurrences:

$$\begin{aligned} x_n &= (x_{n-s} - x_{n-r} - c_n) \bmod b, & (3.11)\\ c_{n+1} &= I(x_{n-s} - x_{n-r} - c_n < 0), & (3.12) \end{aligned}$$

and

$$\begin{aligned} x_n &= (x_{n-r} - x_{n-s} - c_n) \bmod b, & (3.13)\\ c_{n+1} &= I(x_{n-r} - x_{n-s} - c_n < 0), & (3.14) \end{aligned}$$

respectively. Here, c_n is called the *borrow*. For each of those four variants, the maximal period is $M - 1$, where the values of M are $b^r + b^s - 1$, $b^r + b^s + 1$, $b^r - b^s + 1$, and $b^r - b^s - 1$, for AWC, AWC-c, SWB-I, and SWB-II, respectively. Thus, if we choose $b \approx 2^{30}$ and $r = 10$, then the maximal period becomes as large as 2^{300}.

The generators with maximal periods have a nice property that the r-tuples of the consecutive elements of the sequence $x_n, n = 1, 2, \ldots$, produced from these generators are almost equidistributed; that is, the sequence is provably uniform in high dimensions up to r. In order to achieve maximal periods, the parameters (r, s, b) should be chosen so that M is prime and b is a primitive root modulo M. However, since checking the primitivity of an element in a very large prime field is a difficult task in practice, Marsaglia and Zaman [59] listed the parameters for the generators such that the order of b modulo M is very large or near to the maximal, based on careful selection by means of the so-called stringent statistical test. For the following reason, AWC and SWB-I can be viewed as improved variants of the additive and subtractive generators. Suppose $b = 2^w$. Then AWC and SWB-I sequences can have almost the maximum possible period of 2^{rw} attainable by using r w-bit words as an array of memory for the recurrence relation of order r, whereas the periods of the additive and subtractive generators using the same size of memory are at most 2^{r+w-1}, which is far short of the maximum possible. In addition, the AWC/SWB sequences can be generated almost as fast as the additive and subtractive methods.

A random number v_n in $[0,1)$ is constructed by using the consecutive elements of $x_n, n = 1, 2, ...$; that is, for $L > 1$,

$$v_n = \frac{x_{d(n-1)+L}}{b} + \frac{x_{d(n-1)+L-1}}{b^2} + \cdots + \frac{x_{d(n-1)+1}}{b^L}, \tag{3.15}$$

where $d(\geq L)$ is a constant with $\gcd(d, M-1) = 1$. If b is large enough, say around 2^{32}, the random number v_n in $[0,1)$ is constructed by setting $L = 1$, i.e.,

$$v_n = \frac{x_{d(n-1)+1}}{b}. \tag{3.16}$$

Recently, the following result was obtained [97]:

Theorem 3.6 *Let $x_n, n = 1, 2, ...$, be a sequence from AWC, AWC-c, SWB-I, or SWB-II with (r, s, b), for which M is $b^r + b^s - 1$, $b^r + b^s + 1$, $b^r - b^s + 1$, and $b^r - b^s - 1$, respectively. Then the sequence $v_n, n = r+1, r+2, ...$, defined in (3.15) or (3.16) can be viewed as a truncated version of the linear congruential sequence $u_n = X_n/M, n = r+1, r+2, ...$, from $X_n = (b^{-1})^d X_{n-1} \pmod{M}$ with an appropriate X_0, where b^{-1} is the multiplicative inverse of b modulo M.*

Therefore, Marsaglia and Zaman's idea can be viewed as a breakthrough leading to the fast implementation of LCGs with large prime moduli. Figure 3.4 shows the state diagram of a small SWB, where $(x_n, x_{n-1}, x_{n-2}, x_{n-3}, x_{n-4}, c), n = 1, 2, ...$, are produced from the generator, $x_n = x_{n-2} - x_{n-5} - c \pmod{2}$. The number attached to each node is the integer whose binary representation is written as $(x_n, x_{n-1}, x_{n-2}, x_{n-3}, x_{n-4})_2$. If the prime symbol ' is attached to the number, then it means $c = 1$; otherwise, $c = 0$. The number in each bracket shows the corresponding element in the associated linear congruential sequence $X_n = 15X_{n-1} \pmod{29}$, where 15 is the inverse of 2 modulo 29. We observe that the states are divided into two classes, periodic states and transition states, and that the class of periodic states one-to-one corresponds to the associated linear congruential sequence.

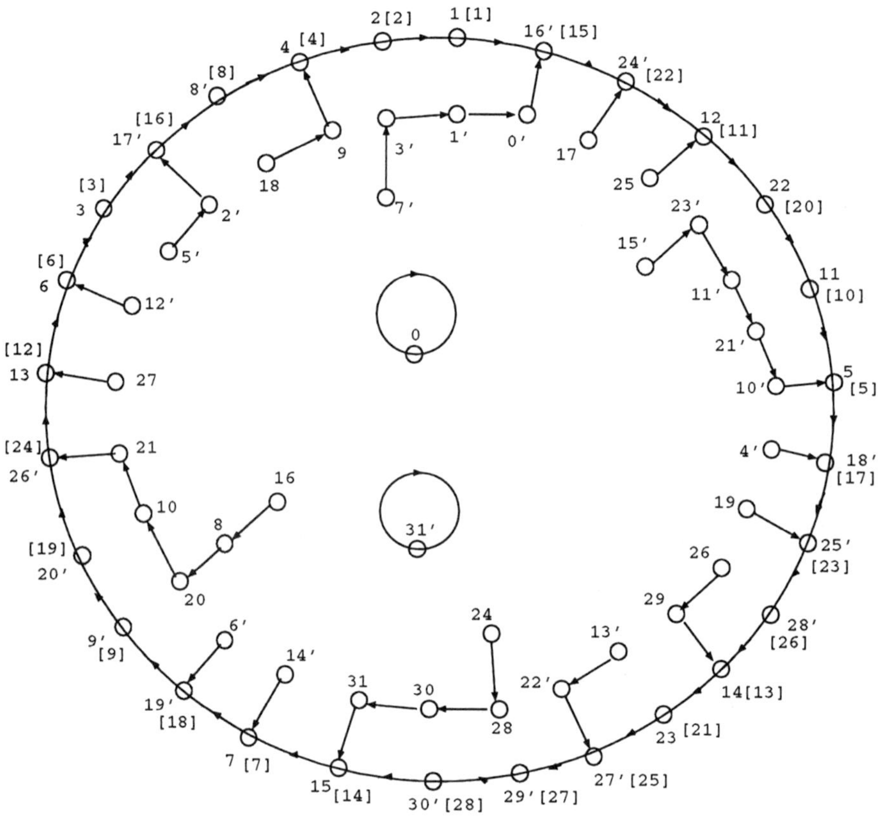

Figure 3.4 State diagram of $(x_n, x_{n-1}, x_{n-2}, x_{n-3}, x_{n-4}, c), n = 1, 2, ...,$ produced by the SWB generator, $x_n = x_{n-2} - x_{n-5} - c \pmod{2}$.

The proof of the above theorem is essentially as follows: Without loss of generality, we consider the AWC sequence. For a nonzero integer X_0, the digit expansion of X_0/M with base b is as follows:

$$\frac{X_0}{M} = \frac{x_1}{b} + \frac{x_2}{b^2} + \cdots,$$

where $M = b^r + b^s - 1$. By multiplying both sides of the equation by M, we notice that the sequence $x_n, n = 1, 2, ...,$ satisfies the recurrence relation (3.7) of the AWC generator in reverse order. Thus, the evolution of the AWC sequence corresponds to

$$\frac{X_{n-1}}{M} = \frac{bX_n}{M} \pmod{1}, \text{ or } \frac{X_n}{M} = \frac{b^{-1}X_{n-1}}{M} \pmod{1},$$

where b^{-1} is the multiplicative inverse of b modulo M. Therefore, it is now sufficient to clarify the maximum length of the transition states, taken over all arbitrary initial states $s_0 = (x_{r-1}, x_{r-2}, ..., x_0, c)$. Hereafter, S denotes the class of all states, and S^* denotes the class of periodic states. The next lemma shows that if $s_0 \neq (b-1, \ldots, b-1, 1)$, the sequence $s_n, n = 0, 1, ...,$ will reach S^* after at most r steps. Note that if $s_0 = (b-1, \ldots, b-1, 1)$, then $s_n = s_0 \notin S^*$ for all n.

Lemma 3.1 *For any $s_0 \in S$ with $s_0 \neq (b-1, \ldots, b-1, 1)$, we have $s_n \in S^*$ for all $n \geq r$.*

Proof: Let $s_0 = (x_{-r+1}, \ldots, x_0, c_1)$ and let

$$\begin{aligned} X_0 &\equiv c_1 + \sum_{j=1}^{r} x_{1-j} b^{r-j} + \sum_{j=1}^{s} x_{1-j} b^{s-j} \\ &= \sum_{j=0}^{r} y_{1-j} b^{r-j}. \end{aligned}$$

Here, we note that $X_0 < M$ because $s_0 \neq (b-1, \ldots, b-1, 1)$. We assume that $s_0 \notin S^*$ and $\bar{s}_0 = (\bar{x}_{-r+1}, \ldots, \bar{x}_0, \bar{c}_1) \in S^*$, where we put $X_0/M = \sum_{n=1}^{\infty} \bar{x}_{1-n} b^{-n}$. Then both $s_r = (x_1, \ldots, x_r, c_{r+1})$ and $\bar{s}_r = (\bar{x}_1, \ldots, \bar{x}_r, \bar{c}_{r+1})$ can be expressed identically in terms of $y_{1-r}, \ldots, y_0, y_1$ in the following way:

$$\begin{aligned} x_1 &= (x_{1-r} + x_{1-s} + c_1) \bmod b = y_{1-r} = \bar{x}_1, \\ &\vdots \end{aligned}$$

$$\begin{aligned}
x_s &= (x_{s-r} + x_0 + c_s) \bmod b \;=\; y_{s-r} = \bar{x}_s, \\
x_{s+1} &= (x_{s-r+1} + x_1 + c_{s+1}) \bmod b \;=\; (y_{s-r+1} + x_1) \bmod b = \bar{x}_{s+1}, \\
c'_{s+2} &= I[y_{s-r+1} + x_1 \geq b] = \bar{c}'_{s+2}, \\
x_{s+2} &= (x_{s-r+2} + x_2 + c_{s+2}) \bmod b \;=\; (y_{s-r+2} + x_2 + c'_{s+2}) \bmod b = \bar{x}_{s+2}, \\
&\;\vdots \\
x_r &= (x_0 + x_{r-s} + c_r) \bmod b \;=\; (y_0 + x_{r-s} + c'_r) \bmod b = \bar{x}_r, \\
c'_{r+1} &= I[y_0 + x_{r-s} + c'_r \geq b] = \bar{c}'_{r+1}, \\
c_{r+1} &= y_1 + c'_{r+1} = \bar{c}_{r+1}.
\end{aligned}$$

Thus, we have $s_r = \bar{s}_r \in S^*$. It is easy to see that once s_r gets in S^*, all $s_n, n > r$ are also in S^*. □

For the case of $L = 1$, we have

$$0 \leq u_n - v_n = \frac{x_{d(n-1)}}{b^2} + \frac{x_{d(n-1)-1}}{b^3} + \cdots \leq \frac{1}{b}.$$

In other words, the distance between v_n and u_n is smaller than the 'resolution' of the generators. Therefore, the sequence $v_n = x_{d(n-1)}/b, n = 1, 2, ...$, in (3.16) is almost equivalent to $u_n, n = 1, 2, ...$, which is a linear congruential sequence from $X_n = b^{-1} X_{n-1} \pmod{M}$. However, we have bad news:

Theorem 3.7 *The lattice structure of all AWC/SWB generators with $L = d = 1$ is always bad in dimensions higher than r.*

Proof: Without loss of generality, we consider only AWC generators. From the definition of the generators, the equation of hyperplanes on which all the $r+1$ dimensional points $(u_n, u_{n+1}, ..., u_{n+r})$, $n = 1, 2, ...$, lie is given by

$$u_n - u_{n-s} - u_{n-r} = j \text{ or } j + \tfrac{1}{b}$$

with $j = -2, -1, 0, 1$. Since b is large enough, $1/b$ is negligible. Therefore, for each j, a twin of hyperplanes corresponding to j and $j + 1/b$ can be regarded approximately as one hyperplane. The distance between such approximate hyperplanes is calculated from the coefficient of the LHS, that is, $(1^2 + (-1)^2 + (-1)^2)^{-1} = \frac{1}{\sqrt{3}} = .577..$, which means that the lattice structure is far from uniform distribution. □

For the case of $L > 1$, we have

$$0 \le u_n - v_n = \frac{x_{d(n-1)}}{b^{L+1}} + \frac{x_{d(n-1)-1}}{b^{L+2}} + \cdots \le \frac{1}{b^L}.$$

Thus, the distance between u_n and v_n is negligible. The implication of this result can be seen in the example in Figure 3.5, which shows the two-dimensional plot of (v_n, v_{n+1}) over the entire period of the sequence generated by

$$v_n = \frac{1}{2^9}\sum_{j=0}^{8} x_{dn+j}2^j$$

with $d = L = 9$, where x_n is generated by $x_n = x_{n-2} - x_{n-9} - c_n \pmod 2$. The period is $2^9 - 2^2 = 508$. Obviously, this sequence is almost the same as that used in Figure 3.1, where the linear congruential sequence is given as

$$X_n = 170X_{n-1} \pmod{509}.$$

When we look very closely at these figures, we can find very small differences due to the truncation. If we choose $d = 55$ with $L = 9$, then we obtain the plot corresponding to the one in Figure 3.2, since $59 = 2^{453} \pmod{509}$ is the inverse of 2^{55} modulo 509, This means that there is a possibility that we can obtain AWC/SWB generators with good lattice structure by carefully choosing a relatively large d compared to L.

EXAMPLE **3.1 [A "Classroom" AWC Generator]**
We examine the "classroom" AWC generator proposed by Marsaglia and Zaman [59]. The sequence is defined by

$$v_n = \frac{1}{6^L}\sum_{j=0}^{L-1} x_{Ln+j}6^j, \tag{3.17}$$

where x_n is generated by $x_n = (x_{n-21} + x_{n-2} + c_n) \bmod 6$ and $L = d$. Since $M = 6^{21} + 6^2 - 1 = 21,936,950,640,377,891$ is prime and $b = 6$ is a primitive root modulo M, the sequence $x_n, n = 1, 2, ...,$ has a period of $M - 1$. There are different values of $L = 1, 3, 7, 9, 11, 17, 19$ satisfying $\gcd(L, M-1) = 1$. The LCG representation is given by

$$X_n = (6^{-1})^L X_{n-1} \bmod M; \qquad u_n = \frac{X_n}{M}.$$

For small L, such as 1 or 3, the resolution is much too low, and as a result, the LCG is not a good approximation of the AWC sequence. We have applied

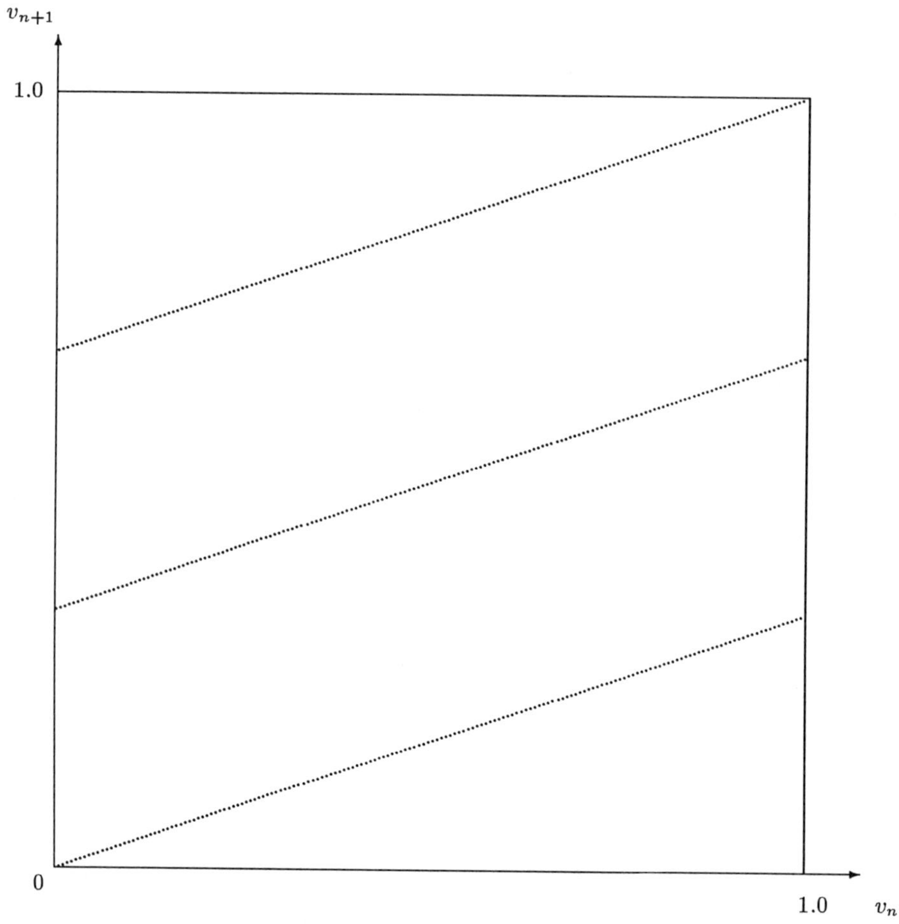

Figure 3.5 A set of two-dimensional points, $(v_n, v_{n+1}), n = 1, ..., 508$, produced by $v_n = \sum_{j=0}^{8} x_{9n+j} 2^{j-9}$, where the subtract-with-borrow generator is given as $x_n = x_{n-2} - x_{n-9} - c_n \pmod 2$ with starting values $(x_1, ..., x_9, c_{10}) = (1, 0, ..., 0)$.

the spectral and Beyer tests to the lattice from the consecutive k elements, i.e., $(j_1, ..., j_k) = (1, ..., k)$, of the corresponding LCGs for the rest of L and $k = 2, ..., 20$. The results are given in Table 3.1, taken from Tezuka et al. [97]. The values ν_k^{-1} and B_k are respectively the distance between hyperplanes in the unit hypercube, and Beyer's ratio, in dimension k. Marsaglia and Zaman said that the x_i's could provide an excellent simulation of independent throws of a dice. If this is true, the sequence from (3.17) should not have any structure. In reality, it turns out that the sequence always has a lattice structure for all values of L; moreover, the lattice structure is very bad, particularly for $L = 17$ and $L = 19$. □

3.3.2 LCGs with Large Composite Moduli

We now consider LCGs in which M is a large composite integer, that is, $M = p_1^{e_1} \cdots p_J^{e_J}$, where $p_j, j = 1, ..., J$, are all distinct primes and $e_j \geq 1$ for $1 \leq j \leq J$. In this case, the maximal period becomes the least common multiple of $\lambda(p_j^{e_j}), j = 1, ..., J$, from Theorem 2.9. Thus the calculation

$$X_n = aX_{n-1} \pmod{M}$$

can be reduced to

$$X_n^{(j)} = aX_{n-1}^{(j)} \pmod{p_j^{e_j}}$$

for $j = 1, ..., J$. Here X_n can be constructed from $X_n^{(j)}, j = 1, ..., J$, by using the Chinese remainder theorem in the following way: Denote $m_j = p_j^{e_j}$. Define $M_j = M/m_j$ and let M_{-j} be an integer such that $M_{-j}M_j \bmod m_j = 1$. Then we have $X_n = \sum_{j=1}^{J} X_n^{(j)} M_{-j} M_j \pmod{M}$, that is,

$$U_n = \frac{X_n}{M} = \sum_{j=1}^{J} \frac{M_{-j} X_n^{(j)}}{m_j} \pmod{1}. \tag{3.18}$$

However, this direct implementation based on the Chinese remainder theorem is rather time-consuming, and thus is not suitable for fast generation of random numbers. A breakthrough was made with the following theorem [46, 108]:

Theorem 3.8 *Let $\bar{X}_n = \sum_{j=1}^{J} X_n^{(j)} M_j$, namely*

$$U_n = \frac{\bar{X}_n}{M} = \sum_{j=1}^{J} \frac{X_n^{(j)}}{m_j} \pmod{1}.$$

Then we have $\bar{X}_n = a\bar{X}_{n-1} \pmod{M}$.

Table 3.1 Beyer and spectral test results

L	7	9	11	17	19
B_2	3.572E-6	4.630E-3	0.167	7.662E-11	1.149E-13
B_3	1.000	2.171E-5	3.473E-6	9.926E-8	4.329E-12
B_4	1.251E-4	2.200E-5	1.216E-4	1.286E-4	1.673E-10
B_5	1.251E-4	4.692E-3	7.293E-4	0.167	6.434E-9
B_6	4.380E-3	4.440E-3	2.552E-2	0.205	2.453E-7
B_7	4.372E-3	0.959	6.143E-2	0.669	9.282E-6
B_8	4.372E-3	0.103	0.473	0.567	3.490E-4
B_9	0.153	0.103	0.550	0.750	1.305E-2
B_{10}	7.088E-2	0.222	0.740	0.477	0.476
B_{11}	7.070E-2	0.229	0.589	0.634	0.562
B_{12}	0.627	0.521	0.861	0.703	0.653
B_{13}	0.358	0.513	0.646	0.870	0.639
B_{14}	0.358	0.536	0.658	0.778	0.729
B_{15}	0.551	0.844	0.613	0.724	0.697
B_{16}	0.439	0.733	0.777	0.663	0.867
B_{17}	0.533	0.761	0.769	0.645	0.800
B_{18}	0.777	0.772	0.854	0.737	0.819
B_{19}	0.700	0.853	0.835	0.778	0.909
B_{20}	0.847	0.816	0.864	0.797	0.829
ν_2^{-1}	3.572E-6	9.923E-8	1.654E-8	7.713E-4	1.992E-2
ν_3^{-1}	3.572E-6	4.570E-3	4.762E-3	7.713E-4	1.992E-2
ν_4^{-1}	2.856E-2	4.570E-3	4.762E-3	7.713E-4	1.992E-2
ν_5^{-1}	2.856E-2	4.570E-3	4.762E-3	7.713E-4	1.992E-2
ν_6^{-1}	2.856E-2	4.570E-3	4.762E-3	3.532E-3	1.992E-2
ν_7^{-1}	2.856E-2	4.570E-3	1.182E-2	4.998E-3	1.992E-2
ν_8^{-1}	2.856E-2	4.486E-2	1.182E-2	1.342E-2	1.992E-2
ν_9^{-1}	2.856E-2	4.486E-2	1.839E-2	1.526E-2	1.992E-2
ν_{10}^{-1}	5.573E-2	4.486E-2	2.243E-2	3.542E-2	1.992E-2
ν_{11}^{-1}	5.573E-2	4.486E-2	3.742E-2	3.542E-2	3.475E-2
ν_{12}^{-1}	5.573E-2	4.486E-2	3.904E-2	4.657E-2	4.608E-2
ν_{13}^{-1}	9.713E-2	6.428E-2	7.715E-2	5.185E-2	5.463E-2
ν_{14}^{-1}	9.713E-2	6.496E-2	7.715E-2	7.727E-2	6.441E-2
ν_{15}^{-1}	9.713E-2	6.652E-2	7.715E-2	7.981E-2	7.125E-2
ν_{16}^{-1}	0.100	9.129E-2	8.220E-2	0.104	8.138E-2
ν_{17}^{-1}	0.100	9.853E-2	9.245E-2	0.104	0.103
ν_{18}^{-1}	0.100	9.853E-2	0.102	0.106	0.103
ν_{19}^{-1}	0.120	0.104	0.109	0.114	0.105
ν_{20}^{-1}	0.120	0.114	0.115	0.123	0.117

Compared to (3.18), in this alternative we need only the modulo-one addition of $X^{(j)}/m_j, 1 \le j \le J$, with no multiplications, which considerably alleviates the implementation cost. This generation scheme can be viewed as the combination of several component LCGs by means of modulo-one addition, where any set of LCGs, even AWC/SWB generators, can be used as components. We recommend that the period lengths of components should be coprime to each other so that the combined sequence has the maximum possible period. Furthermore, we know that even if the component generators all have bad lattice structures, the combined generator can have a good lattice structure. The following example [46] illustrates this claim:

EXAMPLE **3.2** Consider the combined generator of two components: $u^{(1)}$ is with $(a_1, c_1, M_1) = (26756, 0, 2^{31} - 1)$ and $u^{(2)}$ is with $(a_2, c_2, M_2) = (30318, 0, 2145483479)$. $U = u^{(1)} - u^{(2)} \pmod 1$, that is, with

$$(a, c, M) = (341690868154039086 8, 0, 4607390686061167913).$$

Computation of Beyer's ratios shows that in three dimensions two components have bad ratios, 0.0167 and 0.1022, but that the combined generator has a very good ratio of 0.7979. □

* * *

Notes and Remarks

The *matrix generator*, which is suitable for implementation on a vector or parallel supercomputer, are cited in several articles [1, 72]. It should be pointed out that this class of generators can be formulated as the parallel generation of a single MRG sequence with different starting points.

The problem of finding the shortest vector in an integer lattice is believed to be NP-hard [29]. This fact is critical in the design of LCGs with very long periods for applications in the area of large-scale computer simulation, since in such cases we need to apply the spectral and/or Beyer tests of high dimensions, say 20 dimensions or more (often around 100 dimensions), to LCGs with (very) long periods over 2^{60} (often around 2^{1000}). The state-of-the-art implementation of the spectral test reportedly enables us in a reasonable CPU time to investigate LCGs with moduli around 2^{500} in up to 40 dimensions.

The discrepancy of linear congruential sequences was fully analyzed by Niederreiter [40, 68]. The theoretical relations between linear congruential sequences and several randomized (or Monte Carlo) algorithms have been analyzed by Bach [8] and by Traub and Woźniakowski [102].

4

BEYOND LINEAR CONGRUENTIAL GENERATORS

Random number generation is too important
to be left to chance.
– Robert R. Coveyou (1969)

Developing random number generators which involve some nonlinear operation in their generation schemes has been a central research issue in this area, since the nonlinearity has long been believed to be useful to make the resulting sequences look more random. In this chapter, we discuss three types of nonlinear generator. The first and second ones are defined by slightly modifying linear congruential generators, using polynomial arithmetic and multiplicative inversion, respectively. The third type consists of random number generators for cryptographic applications, which require the sequences produced to have a certain property of ' unpredictability.' This property turns out to be strongly connected with the 'nonlinearity' involved in the generation scheme. In fact, linear congruential sequences are known to be unsuitable for such applications because of their polynomial-time 'predictability.'

4.1 LCGS USING POLYNOMIAL ARITHMETIC

We define linear congruential sequences with respect to polynomial arithmetic over finite fields as follows: for $n = 1, 2, ...,$

$$f_n(z) = g(z) f_{n-1}(z) + h(z) \bmod M(z),$$

$$u_n(z) \quad = \quad \frac{f_n(z)}{M(z)} = y_1^{(n)} z^{-1} + y_2^{(n)} z^{-2} + \ldots, \tag{4.1}$$

where $g(z), h(z), M(z)$, and $f_n(z)$ are polynomials in $GF\{q, z\}$, and q is a power of prime p, i.e., $q = p^w$ (w is a positive integer). We denote this generator by the triplet $G = (g, h, M)$ and call it an $LS(q)$ generator. Let $S(z) = \sum_{j=1}^{\infty} c_j z^{-j}$, where $c_j \in GF(q)$. Then we define σ as a mapping from $GF\{q, z\}$ to the real field as follows:

$$\sigma(S(z)) = \sum_{j=1}^{\infty} \lambda(c_j) q^{-j},$$

where λ is a one-to-one mapping from $GF(q)$ to integers in $\{0, 1, ..., q-1\}$, and is the identity mapping if q is prime. We use σ_L, the truncated version of σ, as

$$\sigma_L(S(z)) = \sum_{j=1}^{L} \lambda(c_j) q^{-j}.$$

Hereafter, we denote the truncated $LS(q)$ sequences by $LS_L(q)$ sequences. We should note that the mapping σ is nonlinear, and that the resulting sequences therefore inherit a certain nonlinear structure.

From the practical viewpoint, there are three important cases of $LS_L(q)$ sequences: (1) $q = 2$ and $L > 1$; (2) $q = 2^w (w > 1)$ and $L = 1$; and (3) q is prime and $L = 1$. In the following sections, we discuss the first two cases, since it is easy to see that the third one is equivalent to MRG, which we discussed in the previous chapter. For a more general treatment of $LS_L(q)$ sequences, see Tezuka [93].

$LS_L(2)$ Sequences with the Word-Size L

For practical use, L depends on the size of q so that $q^L \approx 2^{32}$. In the following discussions, we consider the case $q = 2$, which is the most important in practice.

The following theorem confirms the existence of $LS(2)$ sequences with almost maximum periods.

Theorem 4.1 *Let r be the degree of $M(z)$. For $r > 2$, an $LS(2)$ sequence has the maximal period $2^r - 1$ if and only if $M(z)$ is irreducible, $g(z)$ is a primitive root modulo $M(z)$, and $f_0(z) \neq h(z)/(g(z)+1) \pmod{M(z)}$.*

Proof: Let $\bar{f}_n(z) = f_n(z) - f_{n-1}(z)$. Then $\bar{f}_n(z) = g(z)\bar{f}_{n-1}(z) \pmod{M(z)}$. If $M(z)$ is irreducible and $g(z)$ is primitive modulo $M(z)$, and if $\bar{f}_0(z) \neq 0$

$(\text{mod} M(z))$, then $\bar{f}_n(z), n = 1, 2, ...,$ has a period of $2^r - 1$, that is, $f_n(z), n = 1, 2, ...,$ also has a period of $2^r - 1$. If $M(z)$ is irreducible but $g(z)$ is not primitive modulo $M(z)$, then $f_n(z)$ has a period of at most double the multiplicative order of $g(z)$ modulo $M(z)$, which is less than $2^r - 1$, since a factor of $2^r - 1$ is at least three. For reducible $M(z)$, it is sufficient to consider the case $M(z) = m^d(z)$, where $m(z)$ is irreducible and $d > 1$. Let $c = \deg(m)$; then $r = cd$. If $c = 1$ and $d = 2$, then $f_n(z)$ has the maximum period of 4. For example, the generator $(z + 1, 1, z^2)$ produces the maximum period of sequence. If $c = 1$ and $d > 2$, then $f_n(z), n = 1, 2, ...,$ has a period of at most 2^{cd-1}, which is less than $2^{cd} - 1$. When $c > 1$, according to Theorem 2.23, the period of $f_n(z), n = 1, 2, ...,$ is a factor of $2^{c(d-1)+1}(2^c - 1)$, which can never be equal to 2^{cd} or $2^{cd} - 1$. Since $\bar{f}_0(z) \neq 0 \pmod{M(z)}$ means $f_0(z) \neq h(z)/(g(z) + 1) \pmod{M(z)}$, the proof is complete.□

Unlike the conventional LCGs, these generators cannot produce sequences of the maximum length 2^r when $r > 2$. From the above proof, by putting $y_n(z) = f_n(z) + h(z)/(g(z) + 1) \pmod{M(z)}$, we have the homogeneous form

$$y_n(z) = g(z)y_{n-1}(z) \pmod{M(z)},$$

provided that $M(z)$ is irreducible and $g(z) \neq 1$. Therefore, in this section, we restrict ourselves to $LS(2)$ generators with $h(z) = 0$ and $f_0(z) \neq 0$ unless otherwise specified.

Tausworthe sequences [85] and GFSR sequences [50] are originally defined in terms of LFSR sequences, which lead to fast generation and efficient implementation of these sequences. Let $M(z) = z^r + a_{r-1}z^{r-1} + \cdots + a_0$ be a primitive polynomial over $GF(2)$, and let L be the "word-size." Suppose that an LFSR sequence $x_n, n = r, r + 1, ...,$ follows the recurrence relation $x_n = a_{r-1}x_{n-1} + \cdots + a_0x_{n-r} \pmod{2}$. Tausworthe defined the following sequence:

DEFINITION **4.1** A Tausworthe sequence is defined by, for $n = 1, 2, ...,$

$$u_n = \sum_{j=1}^{L} x_{dn+j}2^{-j}, \tag{4.2}$$

where d is a constant with $0 < d < 2^r - 1$ and $\gcd(d, 2^r - 1) = 1$. □

We have an efficient algorithm for generating a Tausworthe sequence whose characteristic polynomial is a primitive trinomial of the form $M(z) = z^r + z^s + 1$,

$s < r/2$, and $0 < d \leq r - s$. Each element of the sequence $u_n, n = 1, 2, \ldots$, is expressed by its leading L bits in this algorithm. This can be implemented easily, in a "portable" language that supports shifting and XOR operations (such as C), if r is not larger than the computer's word-size L. The algorithm is as follows: Let "A << s" and "A >> s" denote s-bit left-shift, and right-shift, of the L-bit word A, respectively.

Algorithm T: implementation of (4.2) when $0 < d \leq r - s$:
Step 0: A and B are L-bit words.
Step 1: B $\Leftarrow$ ((A << s) XOR A) << $(L - r)$
Step 2: A $\Leftarrow$ (A << d) XOR (B >> $(L - d)$)
Step 3: Output A as the leading L bits of u_i, return to Step 1.

Here, the word A should be initialized in advance as $(x_1, \ldots., x_L)$, the leading L-bits of u_0. When $L > r$, we need to extend the initial r-bit pattern $(x_1, \ldots., x_r)$, chosen by the user, into the L-bit pattern $(x_1, \ldots., x_L)$. The following is an initialization algorithm for that purpose:

An initialization algorithm for the case $r < L$:
Step 0: A $= (0, \ldots, 0, x_1, \ldots., x_r)$.
Step 1: B $\Leftarrow$ ((A << s) XOR A)<< $(L - r)$
Step 2: A $\Leftarrow$ (A << $(L - r)$) XOR (B >> r)
Step 3: Output A as $(x_1, \ldots., x_r, x_{r+1}, \ldots, x_L)$.

DEFINITION **4.2** A GFSR sequence is defined by, for $n = 1, 2, \ldots$,

$$u_n = \sum_{j=1}^{L} x_{d_j + n} 2^{-j}, \tag{4.3}$$

where $L \leq r$ and $d_j, j = 1, 2, \ldots, L$, are integers between 1 and $2^r - 1$. □

Lewis and Payne [50] suggested the following special subclass of GFSR sequences:

DEFINITION **4.3** A Lewis-Payne sequence is defined by, for $n = 1, 2, \ldots$,

$$u_n = \sum_{j=1}^{L} x_{dj+n} 2^{-j}, \tag{4.4}$$

where $L \leq r$ and $\gcd(d, 2^r - 1) = 1$. □

Lewis and Payne also employed a primitive trinomial as the characteristic polynomial of $x_n, n = 1, 2, ...$, in order to realize a fast generation scheme for the sequence $u_n, n = 1, 2, ...$, in the following way: Let $M(z) = z^r + z^s + 1$. The sequence can then be generated by the scheme

$$u_n = u_{n-r+s} \text{ XOR } u_{n-r}, \tag{4.5}$$

for which we need to compute initial values $u_1, ..., u_r$ in advance. Table 4.1 gives all the primitive trinomials over $GF(2)$, $z^r + z^s + 1 (s < r/2)$, for up to the 30th Mersenne primes $2^r - 1$ [34].

Now, we show that Tausworthe and Lewis-Payne sequences are special cases of $LS(2)$ sequences.

Proposition 4.1 *Tausworthe sequences can be formulated as $LS_L(2)$ sequences from $G = (g, h, M)$ such that $M(z)$ is the characteristic polynomial for a Tausworthe sequence, that is, a primitive polynomial over $GF(2)$, $g(z)$ is primitive modulo $M(z)$, $h(z) \equiv 0$, and L is the "word-size."*

Proof: Denote $M(z) = z^r + a_{r-1}z^{r-1} + \cdots + a_0$. Since $M(z)$ is a primitive polynomial over $GF(2)$, there exists an integer $d, 0 \leq d < 2^r - 1$, such that $g(z) = z^d \pmod{M(z)}$. Suppose that $f_0(z)/M(z) = x_1 z^{-1} + x_2 z^{-2} + \cdots$. Then $M(z) \times (x_1 z^{-1} + x_2 z^{-2} + \cdots) = f_0(z)$; that is, no fractional terms exist in the LHS. Hence $x_n, n = r+1, r+2, ...$, follows the recurrence relation $x_n = a_{r-1}x_{n-1} + \cdots + a_0 x_{n-r} \pmod 2$. We have

$$\frac{f_{n+1}(z)}{M(z)} = \frac{z^d f_n(z)}{M(z)} \pmod 1.$$

Therefore, an $LS_L(2)$-sequence $u_n, n = 1, 2, ...$, from $G = (g, 0, M)$ is written as

$$u_n = \sum_{j=1}^{L} x_{dn+j} 2^{-j}, \tag{4.6}$$

which is exactly a Tausworthe sequence, since we have $\gcd(d, 2^r - 1) = 1$ from the assumption. Thus the proof is complete. □

Next, we consider a similar formulation for Lewis-Payne sequences. The result implies that Lewis-Payne sequences contain Tausworthe sequences as a proper subclass.

Table 4.1 The primitive trinomials over $GF(2)$, $z^r + z^s + 1 (s < r/2)$, up to the 30th Mersenne prime $2^r - 1$

r	s
2	1
3	1
5	2
7	1, 3
13	none
17	3, 5, 6
19	none
31	3, 6, 7, 13
61	none
89	38
107	none
127	1, 7, 15, 30, 63
521	32, 48, 158, 168
607	105, 147, 273
1279	216, 418
2203	none
2281	715, 915, 1029
3217	67, 576
4253	none
4423	271, 369, 370, 649, 1393, 1419, 2098
9689	84, 471, 1836, 2444, 4187
9941	none
11213	none
19937	881, 7083, 9842
21701	none
23209	1530, 6619, 9739
44497	8575, 21034
86243	none
110503	25230, 53719
132049	7000, 33912, 41469, 52549, 54454

Proposition 4.2 *Lewis-Payne sequences can be formulated as $LS_L(2)$ sequences from $G = (g, h, M)$ such that $M(z)$ is an irreducible polynomial of degree r over $GF(2)$, $g(z)$ is primitive modulo $M(z)$, $h(z) \equiv 0$, and L is the "word-size."*

Proof: Let $F(z)$ be the characteristic polynomial of $x_n, n = 1, 2, \ldots$, in (4.4) and let $M(z)$ be that of the decimated sequence $x_{dn}, n = 1, 2, \ldots$. Note that $F(z)$ is a primitive polynomial, and $M(z)$ is irreducible. Let α be such that $F(\alpha) = 0$, then $x_n = \mathrm{Tr}(\gamma\alpha^n)$ for some γ. Note that $M(\alpha^d) = 0$. Since $M(z)$ is irreducible, α can be written as a linear combination of $1, \alpha^d, \alpha^{2d}, \ldots, \alpha^{(r-1)d}$, that is,

$$\alpha = \sum_{j=0}^{r-1} c_j \alpha^{dj},$$

Thus $x_n = \mathrm{Tr}(\gamma\alpha^n) = \mathrm{Tr}(\gamma\alpha^{n-1}\alpha) = \sum_{j=0}^{r-1} c_j \mathrm{Tr}(\gamma\alpha^{n-1}\alpha^{dj}) = \sum_{j=0}^{r-1} c_j x_{dj+n-1}$. By putting $g(z) = c_{r-1}z^{r-1} + \cdots + c_1 z + c_0$, we complete the proof.□

Note that, in general, GFSR sequences cannot be formulated as $LS_L(2)$ sequences.

$LS_1(2^w)$ Sequences with the Word-Size w

We consider an analogous version of linear congruential sequences in $GF\{q, z\}$ with $q = 2^w$, where w is approximately the word-size of a computer, that is, $w \approx 32$. Then the truncated $LS_L(2^w)$ sequence with $L = 1$ is given as follows: for $n = 1, 2, \ldots$,

$$u_n = \frac{\lambda(y_1^{(n)})}{q},$$

where $y_1^{(n)}$ is the coefficient of z^{-1} in the formal Laurent series $f_n(z)/M(z)$ in (4.1). Consider the case in which $M(z) = z^r - m_{r-1}z^{r-1} - \cdots - m_0$, $g(z) \equiv z$, and $h(z) \equiv 0$. If we use the polynomial representation of elements in $GF(2^w)$ based on an irreducible polynomial $Q(z) = z^w + a_{w-1}z^{w-1} + \cdots + a_0$ over $GF(2)$, writing $y_n(z)$ for $y_1^{(n)}$, and $m_j(z)$ for m_j, $0 \le j < r$, then we arrive at the analogue of multiple recursive generators with respect to polynomial arithmetic over $GF(2)$ as follows:

$$y_n(z) = m_{r-1}(z)y_{n-1}(z) + \cdots + m_0(z)y_{n-r}(z) \quad (\mathrm{mod} Q(z)). \tag{4.7}$$

Note that if $M(z)$ is a primitive polynomial over $GF(2^w)$, then the period of the sequence $y_n(z), n = 1, 2, \ldots$, becomes $2^{wr} - 1$.

How do we implement this efficiently? If we define

$$u_n(z) = \frac{y_n(z)}{Q(z)}, \tag{4.8}$$

then we can exploit the following relation:

$$u_n(z) = m_{r-1}(z)u_{n-1}(z) + \cdots + m_0(z)u_{n-r}(z) \pmod{1}.$$

If $Q(z) = z^w + z^v + 1$, $v < w/2$, and $m_i(z) = z^{d_i} \pmod{Q(z)}$ with $0 < d_i \le w - v$ for $i = 0, ..., w-1$, then we can employ the same implementation method as for Tausworthe sequences by maintaining the array $u_{n-1}(z), ..., u_{n-r}(z)$. Namely, in this case, the leading L bits of each $m_{r-j}(z)u_{n-j}(z), 1 \le j \le r$, can be calculated by using Algorithm T on Page 86.

We give two sets of parameters that have the maximal periods $2^{rw} - 1$: One is $r = 5$, $Q(z) = z^{31} + z^3 + 1$, $m_0(z) = z^{23} \pmod{Q(z)}$, $m_3(z) = z^5 \pmod{Q(z)}$, and $m_i(z) = 0 \pmod{Q(z)}$ for other i's in $0 \le i \le r-1$. The other is $r = 7$, $Q(z) = z^{29} + z^2 + 1$, $m_0(z) = z^{19} \pmod{Q(z)}$, $m_2(z) = z^{16} \pmod{Q(z)}$, and $m_i(z) = 0 \pmod{Q(z)}$ for other i's in $0 \le i \le r-1$.

4.1.1 Lattice Structure of $LS(2)$ Sequences

In the same way as for the linear congruential sequence discussed in the last chapter, we consider the k-tuples $(f_{j_1+n}(z)/M(z), ..., f_{j_k+n}(z)/M(z))$, $n = 1, 2, \ldots$, produced by (4.1), where $j_1, ..., j_k$ are integers with $0 \le j_1 < \cdots < j_k \le 2^r - 1$. These are expressed by the grid (shifted lattice) $\mathcal{L}_k + \lambda$, where $\mathcal{L}_k$ is a (polynomial) lattice in $GF\{2, z\}^k$ with the basis

$$\begin{aligned}
\mathbf{e}_1 &= \frac{1}{M(z)}(1, g(z)^{j_2-j_1}, \ldots, g(z)^{j_k-j_1}),\\
\mathbf{e}_2 &= (0, 1, 0, \ldots, 0),\\
&\vdots\\
\mathbf{e}_k &= (0, 0, 0, \ldots, 1),
\end{aligned}$$

and $\lambda = \frac{h(z)}{M(z)(g(z)-1)}(0, g(z)^{j_2-j_1} - 1, ..., g(z)^{j_k-j_1} - 1)$. Hereafter, we call $\mathcal{L}_k$ the lattice associated with $LS(2)$ sequences. We shall deal with the case $\lambda = (0, ..., 0)$ in the following sections unless otherwise specified, because the uniform distribution property is essentially independent of the value of λ.

In this section, we show that the k-dimensional distribution of $LS(2)$ sequences is very closely related to the successive minima as well as reduced bases of the associated lattice. Before going into details, we need to explain how to measure the k-dimensional uniformity of the sequences. Let

$$R_0 = (\frac{g^{j_1}(z)}{M(z)}, \ldots, \frac{g^{j_k}(z)}{M(z)}),$$

then the lattice $\mathcal{L}_k$ can be represented as the A-submodule of $GF\{2,z\}^k$ defined by

$$\mathcal{L}_k = AR_0 + A^k,$$

where $A = GF[2,z]$. Note that $\mathcal{L}_k$ is cyclic modulo one, since it contains A^k. The lattice points produced from $LS(2)$ sequences over the entire period constitute a subset, denoted by P, of $\mathcal{L}_k \cap C_0^{(k)}$. Note that the subset P does not contain the origin $(0, ..., 0)$. We define an *equidissection* $E_k(l)$ of the k-dimensional unit hypercube into 2^{kl} cubic cells as the set of all cubic cells in $[0,1)^k$ whose sides have a length of 2^{-l} and whose corners have coordinates that are all multiples of 2^{-l}. That is,

$$E_k(l) = \{\prod_{h=1}^{k} [i_h 2^{-l}, (i_h+1)2^{-l}) \mid 0 \le i_1, ..., i_k < 2^l\}.$$

Then define a frequency function $f_l : C_{-l}^{(k)} \to \mathbb{N} \cup \{0\}$ by

$$f_l(i_1, ..., i_k) = |\{(s_1, ..., s_k) \in P \mid (\sigma_l(s_1), ..., \sigma_l(s_k)) \in \prod_{h=1}^{k} [i_h 2^{-l}, (i_h+1)2^{-l})\}|.$$

For each integer m, let

$$\varphi_{l,k}(m) = |\{(i_1, ..., i_k) \mid f_l(i_1, ..., i_k) = m\}|,$$

which represents the number of cells of $E_k(l)$ that contain exactly m points. The following theorem is due to Couture [22]:

Theorem 4.2 *For $1 \le i \le k$, let l_i be the ith successive minimum of the lattice $\mathcal{L}_k$ associated with an $LS(2)$ generator $G = (g, 0, M)$, where $M(z)$ is an irreducible polynomial of degree r over $GF(2)$, and $g(z)$ is primitive modulo $M(z)$. Define the quantity $d(l)$ as follows:*

$$d(l) = \sum_{i=1}^{k} (-l - l_i)^+, \tag{4.9}$$

where $(t)^+ = t$ if $t > 0$, or 0 otherwise. Then we have the following table, which gives the number of cells denoted by $\varphi_{l,k}(m)$ for all values of m for which it could be non-zero.

Table 4.2 Values of $\varphi_{l,k}(m)$ that could be non-zero for a given resolution l

m	$\varphi_{l,k}(m)$
$2^{d(l)}$	$2^{r-d(l)}-1$
$2^{d(l)}-1$	1
0	$2^{lk}-2^{r-d(l)}$

Proof: Let $\mathbf{b}_1, ..., \mathbf{b}_h$ be as in Theorem 2.36. For each nonpositive integer $j \leq 0$, let h_j be the number of points $\mathbf{b}_i$ with $|\mathbf{b}_i| \leq j$ and, for $1 \leq i \leq h_j$, let $\mathbf{b}_i^{(j)} = z^{j-l_i}\mathbf{b}_i$. Then, $|\mathbf{b}_i^{(j)}| = j$. We prove that for any integer $l \leq 0$, the system $\{\mathbf{b}_i^{(j)} | 1 \leq i \leq h_j, j < l\}$ is a basis for $\mathcal{L}_k \cap C_l^{(k)}$ over $GF(2)$. From this, the equation (4.9) easily follows because the set of vectors $\{\mathbf{b}_i^{(j)} | 1 \leq i \leq h_j, l_i \leq j < l\}$ is linearly independent over $GF(2)$.

Without loss of generality, we assume $l_i < l \leq l_{i+1}$. Suppose that there exists a vector $\mathbf{b} \in \mathcal{L}_k \cap C_l^{(k)}$ which cannot be written as a linear combination over $GF(2)$ of the vectors of the system $\{\mathbf{b}_i^{(j)} | 1 \leq i \leq h_j, j < l\}$. Then this vector should be the $(i+1)$th reduced basis, from Theorem 2.36. This is a contradiction. □

Note that the case $m = 0$ does not occur if $lk = r - d(l)$. Now, the following result of Tezuka [89] can be derived as a corollary of Theorem 4.2:

Corollary 4.1 *Each cell of $E_k(l)$ contains exactly 2^{r-lk} points of the lattice $\mathcal{L}_k$ associated with an $LS(2)$ generator $G = (g, 0, M)$ if and only if $l_k \leq -l$, where $M(z)$ is irreducible of degree r over $GF(2)$ and $g(z)$ is primitive modulo $M(z)$.*

This means that the largest successive minimum l_k plays an important role in investigating the uniform distribution of $LS(2)$ sequences. According to Mahler's famous theorem (Theorem 2.39), the ith successive minimum of a polynomial lattice is equal in absolute value to the $(k-i)$th successive minimum of the dual lattice. Hence, the norm of the shortest vector of the dual lattice is exactly equal to the resolution of the sequences. This result corresponds to the conventional spectral test for linear congruential sequences, in which the length of the shortest vector of the dual lattice is equal to the reciprocal of the maximum distance between hyperplanes on which all the points lie. Therefore, we can say that the above corollary defines the spectral test for $LS(2)$ sequences.

EXAMPLE **4.1 [Lattice structure of $LS(2)$ sequences]**
For brevity, we consider the two-dimensional case. Let $\mathbf{b}_1$,$\mathbf{b}_2$ be the reduced basis of the lattice $\mathcal{L}_2$. Since $|\mathbf{b}_1| \leq |\mathbf{b}_2|$, each cell of the equidissection $E_2(-|\mathbf{b}_2| - 1)$ contains $2^{r+2|\mathbf{b}_2|}$ quadrilaterals, each consisting of four lattice points P_1, P_2, P_3, and P_4 specified as $P_2 = P_1 \otimes \sigma(\mathbf{b}_1)$, $P_3 = P_1 \otimes \sigma(\mathbf{b}_2)$, and $P_4 = P_1 \otimes \sigma(\mathbf{b}_1) \otimes \sigma(\mathbf{b}_2)$, where $\otimes$ means the coordinate-wise application of XOR (the bit-wise exclusive-or) operation and $\sigma(\mathbf{b})$ denotes a point $(\sigma(b_x), \sigma(b_y))$ for a vector $\mathbf{b} = (b_x, b_y)$. Theorem 4.2 claims that every cell of the equidissection $E_2(-|\mathbf{b}_2|)$ contains an equal number of points, namely, $2^{r+2|\mathbf{b}_2|}$, and that the point set cannot be evenly distributed into smaller cells of the $E_2(l)$ for $l > -|\mathbf{b}_2|$.

Figure 4.1 shows the point set $(u_n, u_{n+1}), n = 1, 2, ...,$ produced by the $LS(2)$ sequence $u_n = \sum_{j=1}^{6} x_{4n+j} 2^{-j}$, where $x_n, n = 1, 2, ...,$ follow the recurrence relation $x_n = x_{n-5} + x_{n-6} \pmod 2$, and also gives an example of the equidissection $E_2(2)$ of the unit space. Clearly, each cell of $E_2(1)$ contains four quadrilaterals each consisting of four points in the point set, and each cell of $E_2(2)$ contains four points. Furthermore, none of $E_2(l), l > 2$, can evenly divide the point set. Here, $\mathbf{b}_1 = (.0001, .000011)_2 = (1/16, 3/64)$ and $\mathbf{b}_2 = (.000001, .01)_2 = (1/64, 1/4)$. □

The next corollary tells us the importance of the first successive minimum l_1:

Corollary 4.2 *Each cell of $E_k(l)$ contains at most one point of the lattice $\mathcal{L}_k$ associated with an $LS(2)$ generator $G = (g, 0, M)$ if and only if $l_1 \geq -l$, where $M(z)$ is an irreducible polynomial over $GF(2)$ and $g(z)$ is primitive modulo $M(z)$.*

As in the case of linear congruential sequences, we also have two criteria for judging the lattice structure of $LS(2)$ sequences: the spectral test and Beyer's ratio. We should notice that Beyer's and Ripley's ratios are equivalent for $LS(2)$ sequences, from Theorem 2.36. Assume that the associated lattices are obtained from consecutive k elements, i.e., $(j_1, ..., j_k) = (1, ..., k)$. The first criterion (to be minimized), based on the spectral test, is defined as

$$S_{kmax}(G) = \max_{2 \leq k \leq kmax} (l_k + [\frac{r}{k}]),$$

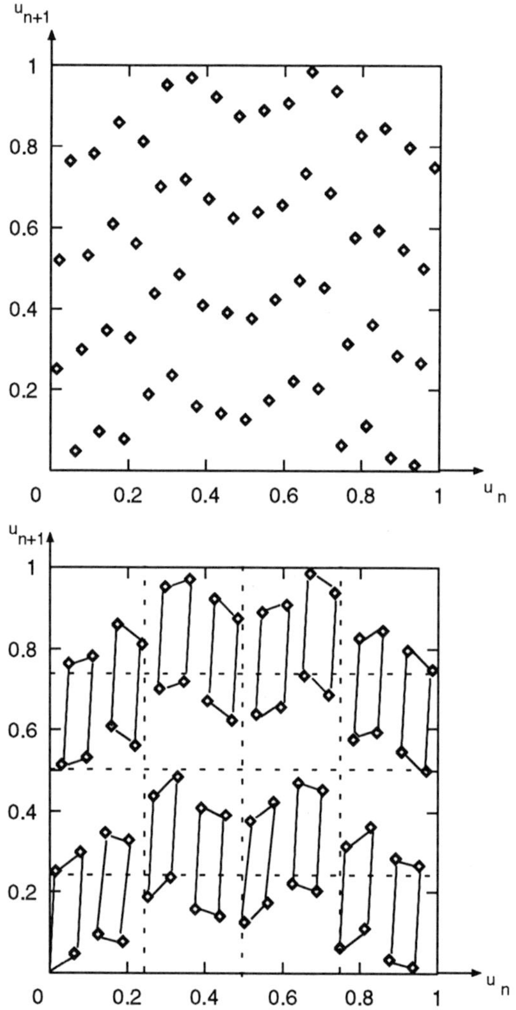

Figure 4.1 (Upper) A set of two-dimensional points, $(u_n, u_{n+1}), n = 1, ..., 63$, produced by the $LS(2)$ sequence, $u_n = \sum_{j=1}^{6} x_{4n+j} 2^{-j}$, where $x_n, n = 1, 2, ...$, follows the recurrence relation $x_n = x_{n-5} + x_{n-6} \pmod 2$. (Lower) Sixteen quadrilaterals from the same point set divided by the equidissection $E_2(2)$.

where l_k is the last successive minimum of the lattice $\mathcal{L}_k$ associated with the generator G and $r = \deg(M)$. Its basic idea is to compare the value of l_k with its theoretical lower bound. The best case is $S_{kmax}(G) = 0$. The other criterion, based on Beyer's ratio, is defined as

$$B_{kmax}(G) = \max_{2 \leq k \leq kmax} (l_k - l_1),$$

where l_1 is the first successive minimum of the lattice $\mathcal{L}_k$ associated with the generator G. It is most desirable that $B_{kmax}(G) \leq 1$.

It is easy to derive the following result from the facts that $-\sum_{i=1}^{k} l_i = r$ and $l_1 \leq l_2 \leq \cdots \leq l_k \leq 0$.

Proposition 4.3 *Let l_1 and l_k be the first and the last successive minimum of the lattice $\mathcal{L}_k$ associated with an $LS(2)$ generator $G = (g, 0, M)$. If $l_k - l_1 \leq 1$, then $l_k = -\lceil r/k \rceil$, where $r = \deg(M)$.*

Therefore, good $LS(2)$ sequences with respect to Beyer's ratio always have the best performance with respect to the spectral test; not vice versa.

EXAMPLE **4.2 [Five Tausworthe Generators with $r = 31$]**
On the basis of the above results, we now examine four Tausworthe generators G_{T_s}'s with primitive trinomials $z^{31} + z^s + 1$, for $s = 3, 6, 7, 13$, respectively, and the "universally optimal" generator G_A, which has been obtained by André et al. [5], specified as $G_A = (z^{31}, 0, z^{31} + z^{29} + z^{27} + z^{26} + z^{20} + z^{16} + 1)$. Table 4.3 gives the values of l_1 and l_k of the associated lattices obtained from the consecutive k elements, i.e., $(j_1, ..., j_k) = (1, ..., k)$, for all dimensions $2 \leq k \leq 15$.

□

EXAMPLE **4.3 [Mullen and Niederreiter's Generator with $r = 64$]**
Here, we look at another "optimal polynomial," suggested in Mullen and Niederreiter [64]. This one has degree 64 and yields the generator $G_M = (z^{64}, 0,\ z^{64} + z^{63} + z^{60} + z^{59} + z^{58} + z^{54} + z^{49} + z^{32} + 1)$. Table 4.4 gives the values of l_1 and l_k of the associated lattices obtained from the consecutive k elements, i.e., $(j_1, ..., j_k) = (1, ..., k)$, for all dimensions $2 \leq k \leq 15$. In dimension $k = 3$, the resolution $l_k = 16$ is much smaller than the maximum resolution 21, and consequently, the uniform distribution is not very good in dimension 3.

Table 4.3 Values of l_1 and l_k in dimensions $k = 2$ to 15 for the generators $G_{Ts}, s = 3, 6, 7, 13$, and G_A

Dim. k		2	3	4	5	6	7	8	9	10	11	12	13	14	15
$[31/k]$		15	10	7	6	5	4	3	3	3	2	2	2	2	2
G_{T3}	$-l_k$	3	3	3	3	3	3	3	3	3	1	1	1	1	1
	$-l_1$	28	25	22	19	16	13	10	7	4	3	3	3	3	3
G_{T6}	$-l_k$	6	6	6	6	1	1	1	1	1	1	1	1	1	1
	$-l_1$	25	19	13	7	6	6	6	6	6	5	5	5	5	5
G_{T7}	$-l_k$	7	7	7	5	3	3	3	3	1	1	1	1	1	1
	$-l_1$	24	17	10	7	7	7	7	4	4	4	4	3	3	3
G_{T13}	$-l_k$	13	5	5	5	3	3	2	2	2	2	2	1	1	1
	$-l_1$	18	13	13	8	8	5	5	5	5	5	3	3	3	3
G_A	$-l_k$	15	10	7	6	5	4	3	3	2	2	2	2	2	2
	$-l_1$	16	11	8	7	6	5	5	4	4	3	3	3	3	3

Table 4.4 Values of l_1 and l_k in dimensions $k = 2$ to 15 for the generator G_M

Dim. k		2	3	4	5	6	7	8	9	10	11	12	13	14	15
$[64/k]$		32	21	16	12	10	9	8	7	6	5	5	4	4	4
G_M	$-l_k$	32	16	16	12	10	8	7	7	6	5	5	4	4	4
	$-l_1$	32	32	16	14	11	10	9	8	7	6	6	6	5	5

□

EXAMPLE **4.4 [Two Hardware Generators with** $r = 127$**]**
Next, we look at two hardware generators, $r = 127$, implemented in the Monte Carlo simulators for Ising spin systems developed by UCSB [74] and by Delft [36] in the 1980's, where $G_U = (z^{24}, 0,\ z^{127} + z^{30} + 1)$ with $L = 24$ for UCSB hardware, and $G_D = (z^{32}, 0,\ z^{127} + z^{15} + 1)$ with $L = 32$ for Delft hardware. The resolutions for the generators G_U and G_D are maximum up to dimension $5 = [\frac{127}{24}]$ and $3 = [\frac{127}{32}]$, respectively. The table gives the values of l_1 and l_k of the associated lattices obtained from the consecutive k elements, i.e., $(j_1, ..., j_k) = (1, ..., k)$, for all dimensions $6 \leq k \leq 19$. In most dimensions, both the generators have bad uniform distributions.

□

EXAMPLE **4.5 [Asymptotically Random Generators with** $r = 607$**]**
Tootill et al. [101] defined an L-bit "asymptotically random" Tausworthe sequence as one that has (1) the maximum resolutions for all dimensions $k \leq [r/L]$ and (2) the prime period $2^r - 1$, and they found a 23-bit "asymptotically ran-

Table 4.5 Values of d in dimensions $k = 6$ to 19 for the generators G_U and G_D

Dim. k		6	7	8	9	10	11	12	13	14	15	16	17	18	19
$\lceil 127/k \rceil$		21	18	15	14	12	11	10	9	9	8	7	7	7	6
G_U	$-l_k$	7	7	7	7	7	7	7	7	7	7	7	6	6	6
	$-l_1$	24	24	24	24	23	17	17	17	17	17	10	10	10	10
G_D	$-l_k$	16	16	15	2	2	2	2	2	2	2	2	2	2	2
	$-l_1$	32	31	16	16	16	16	16	16	16	15	14	14	14	14

dom" Tausworthe sequence $G_{as} = (z^{512}, 0,\ z^{607} + z^{334} + 1)$. By definition, the resolution is maximum for dimensions higher than or equal to $26 = \lceil \frac{607}{23} \rceil$. Table 4.6 gives the values of l_1 and l_k of the associated lattices obtained from the consecutive k elements, i.e., $(j_1, ..., j_k) = (1, ..., k)$, for all dimensions $13 \le k \le 26$. From the table, we confirm that the resolution is not maximum in dimension 25.

Table 4.6 Values of l_1 and l_k in dimensions $k = 13$ to 26 for the generator G_T

Dim. k		13	14	15	16	17	18	19	20	21	22	23	24	25	26
$\lceil 607/k \rceil$		46	43	40	37	35	33	31	30	28	27	26	25	24	23
G_{as}	$-l_k$	44	41	37	36	32	32	31	30	28	27	25	23	23	23
	$-l_1$	51	49	43	39	37	35	33	32	30	29	27	26	25	24

□

Koopman [42] found that the following generator is r-bit asymptotically random: $G_K = (z^{131072}, 0,\ z^{607} + z^{273} + 1)$. Note that this generator as well as the Tootill's one in Example 4.5 can be implemented as GFSR generators with the trinomial recurrence, since $512 = 2^9$ and $131072 = 2^{17}$. Thus, these generators can be said to be the best, at least in terms of their k-distribution and generation speed, although they are not the best in terms of memory usage, because the period is $2^r - 1$, which is much smaller than the maximum possible case $2^{rL} - 1$ obtained by using the same memory area of r L-bit words. But, researchers (e.g. see [99]) have recently become aware of a very discouraging fact that, as will be described below in detail, all of GFSR sequences with trinomial recurrence relations have very bad lattice structures in dimensions higher than r because of their simple generation scheme.

Lattice Structure of Multiple Recursive Sequences

As defined in (4.7), multiple recursive generators with respect to polynomial arithmetic over $GF(2)$ are given as follows:

$$\begin{aligned} y_n(z) &= m_{r-1}(z)y_{n-1}(z) + \cdots + m_0(z)y_{n-r}(z) \quad (\text{mod}Q(z)), \\ u_n(z) &= \frac{y_n(z)}{Q(z)}. \end{aligned}$$

Thus, the lattice structure is exactly the polynomial analogue of the multiple recursive sequence given in (3.3) of Section 3.1.

In particular, GFSR sequences with trinomial recurrence relations,

$$u_{n+r} = u_{n+s} \text{ XOR } u_n,$$

can be written by using polynomial representation as

$$f_{n+r}(z) = f_{n+s}(z) + f_n(z) \quad (\text{mod}Q(z)). \tag{4.10}$$

Thus the dual of the lattice associated with the triples $(f_n(z), f_{n+s}(z), f_{n+r}(z))$, $n = 1, 2, ...,$ is given as

$$\begin{aligned} \mathbf{e}_1^* &= (Q(z), 0, 0), \\ \mathbf{e}_2^* &= (0, Q(z), 0), \\ \mathbf{e}_3^* &= (1, 1, 1). \end{aligned}$$

This means that the nonzero shortest vector of the dual lattice has a length of at most $0 = |\mathbf{e}_3^*|$. Therefore, the resolution of the lattice for these triples from such GFSR sequences is bounded by 0. That is, GFSR sequences with trinomial recurrence relations have an inherently bad lattice structure for the triples $(u_n, u_{n+s}, u_{n+r}), n = 1, 2, ...$ In other words, the third-order correlation with $(j_1, j_2, j_3) = (0, s, r)$ for such GFSR sequences is very high. This bad property is common with all other trinomial-type generators, such as AWC/SWB generators with $L = d = 1$, as well as additive and subtractive generators. In fact, it turned out that all these generators are unsatisfactory for practical applications, such as high-precision Ising model simulations [19, 28, 99].

4.1.2 Combined $LS(2)$ Sequences

As stated in the previous section, simple generation schemes such as GFSR generators with trinomial recurrence relations, in spite of their capability for fast generation of sequences, are not suitable for high-precision Monte Carlo calculations. Some people [21, 96, 105] have recommended that the XOR combination of several GFSR sequences be used in order to avoid such low-order correlations of sequences. Recently, Tezuka and Wong [99] reported that combined $LS(2)$ sequences produced satisfactory results for the same Monte Carlo simulation as Ferrenberg et al. [28] investigated.

DEFINITION **4.4** A combined $LS(2)$ sequence is defined as follows: for $n = 1, 2, \ldots$,

$$U_n = u_n^{(1)} \text{ XOR } \cdots \text{ XOR } u_n^{(J)},$$

where for $j = 1, \ldots, J$, each sequence $u_n^{(j)}, n = 1, 2, \ldots$ is an $LS(2)$ sequence with $(g^{(j)}, 0, M^{(j)})$, and the periods of the sequences, $2^{\deg(M^{(j)})} - 1, j = 1, \ldots, J$, are pairwise coprime. Thus the combined sequence has a period of $\prod_{j=1}^{J}(2^{\deg(M^{(j)})} - 1)$. □

We have the following result: For each $j = 1, \ldots, J$, we define $M_j(z) = M(z)/M^{(j)}(z)$ and let $M_{-j}(z)$ be a polynomial such that $M_{-j}(z)M_j(z) = 1 \pmod{M^{(j)}(z)}$.

Proposition 4.4 *A combined $LS(2)$ sequence can be formulated as an $LS(2)$ sequence from $G = (g, h, M)$, where $h(z) \equiv 0$, $M(z) = M^{(1)}(z) \cdots M^{(J)}(z)$, and*

$$g(z) = \sum_{j=1}^{J} g^{(j)}(z) M_{-j}(z) M_j(z).$$

By using Algorithm T on Page 86 to implement each component $LS(2)$ sequence, we can construct a combined $LS(2)$ sequence quickly. Next, we have

DEFINITION **4.5** A combined multiple recursive sequence is defined as follows: for $n = 1, 2, \ldots$,

$$U_n = u_n^{(1)} \text{ XOR } \cdots \text{ XOR } u_n^{(J)},$$

where for $j = 1, \ldots, J$, each sequence $u_n^{(j)}, n = 1, 2, \ldots$, is a multiple recursive sequence in (4.7) with period $2^{r_j \deg(Q^{(j)})} - 1$, and the periods of the sequences are pair-wise coprime. Thus the period of the combined sequence becomes $\prod_{j=1}^{J}(2^{r_j \deg(Q^{(j)})} - 1)$. □

For $j = 1, \ldots, J$, let $\mathcal{L}_k^{(j)}$ be the lattice in $GF\{2,z\}^k$ associated with each $LS(2)$ sequence $u_n^{(j)}, n = 1, 2, \ldots$. The lattice $\mathcal{L}_k$ associated with the combined sequence $U_n, n = 1, 2, \ldots$, becomes

$$\mathcal{L}_k \cap C_0^{(k)} = (\mathcal{L}_k^{(1)} \cap C_0^{(k)}) \oplus \cdots \oplus (\mathcal{L}_k^{(J)} \cap C_0^{(k)}) \tag{4.11}$$

(direct sum of vector spaces over $GF(2)$). We define $V_j = \mathcal{L}_k^{(j)} \cap C_0^{(k)}$ for $j = 1, \ldots, J$, and also define $V_\Psi = \bigoplus_{j \in \Psi} V_j$, $W_\Psi = V_\Psi \cap C_{-l}^{(k)}$, and $d_\Psi = \dim(W_\Psi)$ for each subset Ψ of $\{1, \ldots, J\}$. If $\Psi = \{1, \ldots, J\}$, we also write $\mathcal{L}_k^{(\Psi)}$, V_Ψ, and W_Ψ as $\mathcal{L}_k$, V, and W respectively. We should notice that all the objects and quantities defined above depend implicitly on k and l. Each d_Ψ can be computed by using Theorem 4.2. Let P denote the set of lattice points in $\mathcal{L}_k \cap C_0^{(k)}$ associated with the combined $LS(2)$ sequence. Then

$$P = V \setminus \bigcup_{|\Psi| = J-1} V_\Psi.$$

Let l_k be the last successive minimum of the lattice $\mathcal{L}_k$ associated with the combined generator $G = (g, 0, M)$. For each i, j with $i \neq j$ in $\{1, \ldots, J\}$, define $M_{ij}(z) = M(z)/(M^{(i)}(z)M^{(j)}(z))$, let $M_{-ij}(z)$ be a polynomial such that $M_{-ij}(z)M_{ij}(z) = 1 \pmod{M^{(j)}(z)}$, for $1 \leq i \neq j \leq J$, and let $g_j(z) = \sum_{i=1,\, i \neq j}^{J} g^{(j)}(z) M_{-ij}(z) M_{ij}(z)$. Denote $\mathcal{L}_k^{(\Psi - \{j\})}$ by $\mathcal{L}_k^{(-j)}$. Let $l_1^{(-j)}$ be the first successive minimum of the lattice $\mathcal{L}_k^{(-j)}$ associated with the generator $(g_j, 0, M_j)$. Let $r = r_1 + \cdots + r_J$ be the degree of $M(z)$. The following theorem was proved by Tezuka [96]:

Theorem 4.3 *If $\min_{j=1}^{J} l_1^{(-j)} \geq l_k$, then for l with $\min_{j=1}^{J} l_1^{(-j)} \geq l \geq l_k$, the number of points $(U_{n+j_1}, \ldots, U_{n+j_k}), n = 1, 2, \ldots$, that are contained in each cell of the equidissection $E_k(-l)$ is equal to an integer in the range $[2^{r+kl} - J, 2^{r+kl}]$.*

Proof: The lattice $\mathcal{L}_k$ associated with the combined sequence has 2^r lattice points, including the origin, and each cell of $E_k(-l_k)$ contains 2^{r+kl_k} lattice points. Note that the lattice points not appearing as $(U_{n+j_1}, \ldots, U_{n+j_k})$ from the combined sequence are contained in at least one of the lattices, $\mathcal{L}_k^{(-j)}, j = 1, \ldots, J$. Since $l_1^{(-j)} \geq l_k, j = 1, \ldots, J$, each of the lattices, $\mathcal{L}_k^{(-j)}, j = 1, \ldots, J$, contains at most one lattice point in every cell of the equidissection $E_k(-l)$ for l with $\min_{j=1}^{J} l_1^{(-j)} \geq l \geq l_k$. It is possible that all of the lattices, $\mathcal{L}_k^{(-j)}, j = 1, \ldots, J$, share one cell of the equidissection $E_k(-l)$ for their distinct lattice points. Therefore, such a cell contains at least $2^{r+kl} - J$ points from the sequence. Thus the proof is complete. □

Roughly speaking, the smaller the (negative) value of l_k, the better the k-dimensional distribution of the combined sequence.

EXAMPLE **4.6 [A Combined $LS(2)$ Generator in SUPER-DUPER]** Our next example is a combined $LS(2)$ generator, which is itself a component of the generator SUPER-DUPER proposed by Marsaglia [56]. This generator is given by $g(z) = (z^{32}, 0, z^{32} + z^{15} + 1)$. Note that $M(z) = z^{32} + z^{15} + 1$ is not irreducible and can be factored as $M(z) = (z^{21} + z^{19} + z^{15} + z^{13} + z^{12} + z^{10} + z^9 + z^8 + z^7 + z^6 + z^4 + z^2 + 1)(z^{11} + z^9 + z^7 + z^2 + 1)$. Therefore, this generator is regarded as a combined $LS(2)$ generator, whose components are $g_1 = (z^{32}, 0, z^{21} + z^{19} + z^{15} + z^{13} + z^{12} + z^{10} + z^9 + z^8 + z^7 + z^6 + z^4 + z^2 + 1)$ and $g_2 = (z^{32}, 0, z^{11} + z^9 + z^7 + z^2 + 1)$. The maximum possible period is $(2^{21} - 1)(2^{11} - 1)$ and therefore almost all initial values give the maximum period.

The results are given in Table 4.7, which gives the values of l_1 and l_k of the associated lattices obtained from the consecutive k elements, i.e., $(j_1, ..., j_k) = (1, ..., k)$, for all dimensions $2 \leq k \leq 15$. The bad behavior of uniform distribution is to be expected in dimensions 3 and more.

Table 4.7 Values of l_1 and l_k in dimensions $k = 2$ to 15 for the components of SUPER-DUPER

Dim. k		2	3	4	5	6	7	8	9	10	11	12	13	14	15
$\lceil 32/k \rceil$		16	10	8	6	5	4	4	3	3	2	2	2	2	2
g	$-l_k$	15	2	2	2	2	2	2	2	2	2	2	2	2	2
	$-l_1$	17	15	15	13	13	11	11	9	9	7	7	5	5	3
g_1	$-l_k$	10	2	2	2	2	2	2	2	2	1	1	1	1	1
	$-l_1$	11	10	9	8	7	6	5	4	3	2	2	2	2	2
g_2	$-l_k$	5	2	2	2	1	1	1	1	1	1	0	0	0	0
	$-l_1$	6	5	4	3	2	2	2	2	2	1	1	1	1	1

□

EXAMPLE **4.7 [A Combined $LS(2)$ Generator of Three Components]** We found a combined $LS(2)$ generator with three components; they have a very nice property:

$$B_{20}(G) = \max_{2 \leq k \leq 20} (l_k - l_1) = 1.$$

The specification of the three components are as follows:

$$\begin{aligned} G_1 &= (z^{13}, 0, z^{28} + z^9 + 1), & (4.12)\\ G_2 &= (z^{20}, 0, z^{29} + z^2 + 1), & (4.13)\\ G_3 &= (z^{17}, 0, z^{31} + z^6 + 1). & (4.14) \end{aligned}$$

From Table 4.8 and Theorem 4.3, the number of points in each cell of the equidissection $E_k(-l_k)$ is equal to an integer in the range $[2^{88+kl_k} - 3, 2^{88+kl_k}]$, where G_{ij} means the combination of G_i and G_j for $1 \le i, j \le 3$, and G is the combination G_1, G_2, and G_3.

A sample C program of this combined Tausworthe generator G is given in Figure A.1 of Appendix A.

Table 4.8 Values of l_1 and l_k in dimensions $k = 2$ to 15 for the generators $G_1, G_2, G_3, G_{12}, G_{23}, G_{13}$, and G

Dim. k		2	3	4	5	6	7	8	9	10	11	12	13	14	15
G_1	$-l_k$	13	6	6	5	2	2	2	2	2	2	2	2	1	1
	$-l_1$	15	13	9	6	6	6	5	5	4	4	4	3	3	2
G_2	$-l_k$	9	9	6	4	4	3	3	2	2	2	2	2	2	1
	$-l_1$	20	11	9	7	6	5	5	5	4	3	3	3	3	2
G_3	$-l_k$	14	8	7	4	4	4	3	3	2	2	2	2	2	2
	$-l_1$	17	14	8	8	7	5	5	4	4	3	3	3	3	3
G_{12}	$-l_k$	28	18	13	11	9	8	7	5	5	5	4	4	4	3
	$-l_1$	29	20	15	12	10	9	8	7	6	6	5	5	5	4
G_{23}	$-l_k$	30	20	15	12	10	8	7	6	5	5	4	4	3	3
	$-l_1$	30	20	15	12	10	9	8	7	7	6	6	5	5	5
G_{13}	$-l_k$	29	17	14	11	9	8	7	6	5	5	4	4	4	3
	$-l_1$	30	22	15	12	10	9	8	7	7	6	6	5	5	4
G	$-l_k$	44	29	22	17	14	12	11	9	8	8	7	6	6	5
	$-l_1$	44	30	22	18	15	13	11	10	9	8	8	7	7	6

Example 4.8 **[A Combination of Two Multiple Recursive Generators]** Our next example is a combination of two multiple recursive generators given on Page 90. The component generators, M_1 and M_2, are formulated as follows: One sequence $\{u_i^{(1)}\}$ is defined by

$$\begin{aligned} f_n^{(1)}(z) &= z^5 f_{n-2}^{(1)}(z) + z^{23} f_{n-5}^{(1)}(z) \pmod{Q^{(1)}(z)}, & (4.15)\\ u_n^{(1)}(z) &= \frac{f_n^{(1)}(z)}{Q^{(1)}(z)}, & (4.16) \end{aligned}$$

where $Q^{(1)}(z) = z^{31} + z^3 + 1$. The other sequence $\{u_i^{(2)}\}$ is

$$f_n^{(2)}(z) = z^{16} f_{n-5}^{(2)}(z) + z^{19} f_{n-7}^{(2)}(z) \pmod{Q^{(2)}(z)}, \tag{4.17}$$

$$u_n^{(2)}(z) = \frac{f_n^{(2)}(z)}{Q^{(2)}(z)}, \tag{4.18}$$

where $Q^{(2)}(z) = z^{29} + z^2 + 1$. Hence, the combined generator M is obtained by

$$U_n = u_n^{(1)} \text{ XOR } u_n^{(2)},$$

for $n = 1, 2, \ldots$. Since the periods of two component generators are coprime, the resulting period becomes $(2^{155} - 1)(2^{203} - 1)$. From Table 4.9, the uniform distribution of the combined sequence is very good in all dimensions $k = 13$ to 40.

A sample C program of this combined MRG generator M is given in Figure A.2 of Appendix A. □

Table 4.9 Values of l_1 and l_k in dimensions $k = 13$ to 40 for the combined and component generators

Dim. k		13	14	15	16	17	18	19	20	21	22	23	24	25	26
$\lfloor 358/k \rfloor$		27	25	23	22	21	19	18	17	17	16	15	14	14	13
M	$-l_k$	26	25	23	22	20	19	18	17	17	16	15	14	14	13
	$-l_1$	28	26	24	23	22	20	19	18	18	17	16	15	15	14
$\lfloor 155/k \rfloor$		11	11	10	9	9	8	8	7	7	7	6	6	6	5
M_1	$-l_k$	10	10	8	8	8	8	8	7	7	7	6	6	6	5
	$-l_1$	13	12	11	10	10	10	9	8	8	8	7	7	7	6
$\lfloor 203/k \rfloor$		15	14	13	12	11	11	10	10	9	9	8	8	8	7
M_2	$-l_k$	12	10	10	10	10	10	10	9	9	8	8	8	8	7
	$-l_1$	20	19	19	19	19	19	12	11	11	10	10	9	9	8

Dim. k		27	28	29	30	31	32	33	34	35	36	37	38	39	40
$\lfloor 358/k \rfloor$		13	12	12	11	11	11	10	10	10	9	9	9	9	8
M	$-l_k$	13	12	12	11	11	11	10	10	10	9	9	9	9	8
	$-l_1$	14	13	13	12	12	12	11	11	11	10	10	10	10	9
$\lfloor 155/k \rfloor$		5	5	5	5	5	4	4	4	4	4	4	4	3	3
M_1	$-l_k$	5	5	5	5	4	4	4	4	4	4	4	4	3	3
	$-l_1$	6	6	6	6	6	5	5	5	5	5	5	5	5	4
$\lfloor 203/k \rfloor$		7	7	7	6	6	6	6	5	5	5	5	5	5	5
M_2	$-l_k$	7	7	6	6	6	6	6	5	5	5	5	5	5	5
	$-l_1$	8	8	8	7	7	7	7	7	6	6	6	6	6	6

Table 4.10 Values of $\varphi_{l,k}(m)$ that could be non-zero for $J = 2$

m	$\varphi_{l,k}(m)$
2^d	$2^{r-d} + 2^{d-d_1-d_2} - 2^{r_1-d_1} - 2^{r_2-d_2}$
$2^d - 2^{d_1}$	$2^{r_1-d_1} - 2^{d-d_1-d_2}$
$2^d - 2^{d_2}$	$2^{r_2-d_2} - 2^{d-d_1-d_2}$
$2^d - 2^{d_1} - 2^{d_2}$	$2^{d-d_1-d_2} - 1$
$2^d - 2^{d_1} - 2^{d_2} + 1$	1
0	$2^{lk} - 2^{r-d}$

Tables of $\varphi_{l,k}(m)$ *for* $J = 2$ *and* 3

In the cases $J = 2$ and 3, Couture [22] obtained tables for the values of $\varphi_{l,k}(m)$ that could be non-zero. The results for $J = 2$ are summarized in Table 4.10, where the first column gives all the possible values of m for which $\varphi_{l,k}(m)$ is not always zero.

For $J = 3$, the results are shown in Table 4.11. The parameter D is defined as follows: Let $V = V_1 + V_2 + V_3$ be a direct sum of vector spaces and $W \subset V$ a subspace. For each $i \neq j$, let $W_i = W \cap V_i$, $W_{ij} = W \cap (V_i + V_j)$, $d = \dim(W)$, $d_i = \dim(W_i)$, and $d_{ij} = \dim(W_{ij})$. Then we define $D = \dim(((V_1 + W) \cap (V_2 + W) \cap (V_3 + W))/W)$. The next example shows that knowing d, d_i, and d_{ij} is not sufficient in general to compute D, where we should note that $D = \dim(V_1 \cap (V_2 + W_{12}) \cap (V_3 + W_{13}))$ if $W_1 + W_2 + W_3 = \{0\}$ (see Couture [22]).

EXAMPLE **4.9** For $i = 1, 2, 3$, let $\dim(V_i) = 2$ and let $\{\mathbf{v}_i, \mathbf{v}_i'\}$ be a basis for V_i. We consider two cases. In the first case, suppose that $W = \mathbb{F}_2 \cdot (\mathbf{v}_1 + \mathbf{v}_2') + \mathbb{F}_2 \cdot (\mathbf{v}_2 + \mathbf{v}_3') + \mathbb{F}_2 \cdot (\mathbf{v}_3 + \mathbf{v}_1')$, where $\mathbb{F}_2 \cdot \mathbf{v}$ means the space $\{0, \mathbf{v}\}$. Then, $W_i = W \cap V_i = \{0\}$ for each i and $H_1 = V_1 \cap (V_2 + W_{12}) \cap (V_3 + W_{13}) = V_1 \cap (V_2 + \mathbb{F}_2 \cdot (\mathbf{v}_1 + \mathbf{v}_2')) \cap (V_3 + \mathbb{F}_2(\mathbf{v}_3 + \mathbf{v}_1')) = V_1 \cap (V_2 + \mathbb{F}_2 \cdot \mathbf{v}_1) \cap (V_3 + \mathbb{F}_2 \cdot \mathbf{v}_1') = (\mathbb{F}_2 \cdot \mathbf{v}_1) \cap (\mathbb{F}_2 \cdot \mathbf{v}_1') = \{0\}$, so that $D = \dim(H_1) = 0$. In the second case, suppose that $W = \mathbb{F}_2 \cdot (\mathbf{v}_1 + \mathbf{v}_2) + \mathbb{F}_2 \cdot (\mathbf{v}_2 + \mathbf{v}_3) + \mathbb{F}_2 \cdot (\mathbf{v}_1' + \mathbf{v}_2' + \mathbf{v}_3')$. Then, $W \cap V_i = \{0\}$ for each i and $H_1 = V_1 \cap (V_2 + W_{12}) \cap (V_3 + W_{13}) = V_1 \cap (V_2 + \mathbb{F}_2 \cdot (\mathbf{v}_1 + \mathbf{v}_2)) \cap (V_3 + \mathbb{F}_2(\mathbf{v}_1 + \mathbf{v}_3)) = V_1 \cap (V_2 + \mathbb{F}_2 \cdot \mathbf{v}_1) \cap (V_3 + \mathbb{F}_2 \cdot \mathbf{v}_1) = \mathbb{F}_2 \mathbf{v}_1$, so that $D = \dim(H_1) = 1$. In both cases, $d = 3$, $d_{ij} = 1$, and $d_i = 0$, but the two cases have different values of D. □

The next result, due to Couture [22], allows us to compute D efficiently.

Table 4.11 Possible signatures and frequencies for generators with three components

m	$\varphi_{l,k}(m)$
2^d	$2^{r-d} + 2^{r_1+d-d_{12}-d_{13}} + 2^{r_2+d-d_{12}-d_{23}} + 2^{r_3+d-d_{13}-d_{23}}$ $-2^{2d-d_{12}-d_{13}-d_{23}} - 2^{r_{12}-d_{12}} - 2^{r_{13}-d_{13}} - 2^{r_{23}-d_{23}}$
$2^d - 2^{d_{12}}$	$2^{r_{12}-d_{12}} - 2^{r_1+d-d_{12}-d_{13}} - 2^{r_2+d-d_{12}-d_{23}}$ $+2^{2d-d_{12}-d_{13}-d_{23}}$
$2^d - 2^{d_{13}}$	$2^{r_{13}-d_{13}} - 2^{r_1+d-d_{12}-d_{13}} - 2^{r_3+d-d_{13}-d_{23}}$ $+2^{2d-d_{12}-d_{13}-d_{23}}$
$2^d - 2^{d_{23}}$	$2^{r_{23}-d_{23}} - 2^{r_2+d-d_{23}-d_{12}} - 2^{r_3+d-d_{13}-d_{23}}$ $+2^{2d-d_{12}-d_{13}-d_{23}}$
$2^d - 2^{d_{12}} - 2^{d_{13}}$	$2^{r_1+d-d_{12}-d_{13}} - 2^{2d-d_{12}-d_{13}-d_{23}}$ $-2^{r_1-d_1} + 2^{d-d_1-d_{23}}$
$2^d - 2^{d_{12}} - 2^{d_{23}}$	$2^{r_2+d-d_{23}-d_{12}} - 2^{2d-d_{12}-d_{13}-d_{23}}$ $-2^{r_2-d_2} + 2^{d-d_2-d_{13}}$
$2^d - 2^{d_{13}} - 2^{d_{23}}$	$2^{r_3+d-d_{13}-d_{23}} - 2^{2d-d_{12}-d_{13}-d_{23}}$ $-2^{r_3-d_3} + 2^{d-d_3-d_{13}}$
$2^d - 2^{d_{12}} - 2^{d_{13}} - 2^{d_{23}}$	$2^{2d-d_{12}-d_{13}-d_{23}} - 2^{d-d_1-d_{23}} - 2^{d-d_2-d_{13}} - 2^{d-d_3-d_{13}}$ $+2^{d_{13}-d_1-d_3} + 2^{d_{12}-d_1-d_2} + 2^{d_{23}-d_2-d_3} - 2^D$
$2^d - 2^{d_{12}} - 2^{d_{13}} + 2^{d_1}$	$2^{r_1-d_1} - 2^{d-d_1-d_{23}}$
$2^d - 2^{d_{12}} - 2^{d_{23}} + 2^{d_2}$	$2^{r_2-d_2} - 2^{d-d_2-d_{13}}$
$2^d - 2^{d_{13}} - 2^{d_{23}} + 2^{d_3}$	$2^{r_3-d_3} - 2^{d-d_3-d_{12}}$
$2^d - 2^{d_{12}} - 2^{d_{13}} - 2^{d_{23}}$ $+2^{d_1}$	$2^{d-d_1-d_{23}} - 2^{d_{13}-d_1-d_3} - 2^{d_{12}-d_1-d_2} + 2^D$
$2^d - 2^{d_{12}} - 2^{d_{13}} - 2^{d_{23}}$ $+2^{d_2}$	$2^{d-d_2-d_{13}} - 2^{d_{12}-d_1-d_2} - 2^{d_{23}-d_2-d_3} + 2^D$
$2^d - 2^{d_{12}} - 2^{d_{13}} - 2^{d_{23}}$ $+2^{d_3}$	$2^{d-d_3-d_{12}} - 2^{d_{23}-d_2-d_3} - 2^{d_{13}-d_1-d_3} + 2^D$
$2^d - 2^{d_{12}} - 2^{d_{13}} - 2^{d_{23}}$ $+2^{d_1} + 2^{d_2}$	$2^{d_{12}-d_1-d_2} - 2^D$
$2^d - 2^{d_{12}} - 2^{d_{13}} - 2^{d_{23}}$ $+2^{d_1} + 2^{d_3}$	$2^{d_{13}-d_1-d_3} - 2^D$
$2^d - 2^{d_{12}} - 2^{d_{13}} - 2^{d_{23}}$ $+2^{d_2} + 2^{d_3}$	$2^{d_{23}-d_2-d_3} - 2^D$
$2^d - 2^{d_{12}} - 2^{d_{13}} - 2^{d_{23}}$ $+2^{d_1} + 2^{d_2} + 2^{d_3}$	$2^D - 1$
$2^d - 2^{d_{12}} - 2^{d_{13}} - 2^{d_{23}}$ $+2^{d_1} + 2^{d_2} + 2^{d_3} - 1$	1
0	$2^{lk} - 2^{r-d}$

Proposition 4.5 *Consider the lattice* $\mathcal{L}'_{3k} = \mathcal{L}_k^{(12)} \times \mathcal{L}_k^{(13)} \times \mathcal{L}_k^{(23)} \subset GF\{2,z\}^{3k}$ *and the mapping* $\eta : GF\{2,z\}^{3k} \mapsto GF\{2,z\}^k$ *defined by* $\eta(\mathbf{v}_1, \mathbf{v}_2, \mathbf{v}_3) = \mathbf{v}_1 + \mathbf{v}_2 + \mathbf{v}_3$. *Let* $\bar{\mathcal{L}}_{3k} = \mathcal{L}'_{3k} \cap \ker(\eta) = \{\mathbf{v} \in \mathcal{L}'_{3k} \mid \eta(\mathbf{v}) = 0\}$. *We then have* $D = \dim(\bar{\mathcal{L}}_{3k} \cap C_{-l}^{(3k)}) - d_1 - d_2 - d_3$.

Before using this result, we have to construct a generating system for $\bar{\mathcal{L}}_{3k}$ in the following way: An element of $\mathcal{L}'_{3k}$ can be written as $\mathbf{v} = (\mathbf{v}_1 + \mathbf{v}_2, \mathbf{v}'_1 + \mathbf{v}'_3, \mathbf{v}''_2 + \mathbf{v}''_3)$, with $\mathbf{v}_1, \mathbf{v}'_1 \in \mathcal{L}_k^{(1)}, \mathbf{v}_2, \mathbf{v}''_2 \in \mathcal{L}_k^{(2)}$ and $\mathbf{v}'_3, \mathbf{v}''_3 \in \mathcal{L}_k^{(3)}$. Such a $\mathbf{v}$ belongs to $\bar{\mathcal{L}}_{3k}$ if and only if

$$\mathbf{v}_1 + \mathbf{v}'_1 + \mathbf{v}_2 + \mathbf{v}''_2 + \mathbf{v}'_3 + \mathbf{v}''_3 = 0. \tag{4.19}$$

If we restrict ourselves to $C_0^{(k)}$ (or modulo one), then $\mathcal{L}_k$ is the direct sum of $\mathcal{L}_k^{(1)}$, $\mathcal{L}_k^{(2)}$, and $\mathcal{L}_k^{(3)}$, from (4.11). Thus, we obtain from (4.19), $\mathbf{v}_1 + \mathbf{v}'_1 = \mathbf{v}_2 + \mathbf{v}''_2 = \mathbf{v}'_3 + \mathbf{v}''_3 = 0$ modulo one, and $\mathbf{v}$ can be written as $(\mathbf{v}_1 + \mathbf{v}_2, -\mathbf{v}_1 + \mathbf{v}_3, -\mathbf{v}_2 - \mathbf{v}_3) = (\mathbf{v}_1, -\mathbf{v}_1, 0) + (\mathbf{v}_2, 0, -\mathbf{v}_2) + (0, \mathbf{v}_3, -\mathbf{v}_3)$ modulo one. Therefore, a generating system for $\bar{\mathcal{L}}_{3k}$ is obtained as the union of $\{(\mathbf{v}_1, -\mathbf{v}_1, 0)\}, \{(\mathbf{v}_2, 0, -\mathbf{v}_2)\}$, $\{(0, \mathbf{v}_3, -\mathbf{v}_3)\}$, and a basis for $A^{3k} \cap \ker(\eta)$, where $\mathbf{v}_1, \mathbf{v}_2$, and $\mathbf{v}_3$ run through the basis of $\mathcal{L}_k^{(1)}, \mathcal{L}_k^{(2)}$, and $\mathcal{L}_k^{(3)}$, respectively. A basis for $A^{3k} \cap \ker(\eta)$ is given by $\{\mathbf{e}_i - \mathbf{e}_{i+k}, \mathbf{e}_i - \mathbf{e}_{i+2k} \mid i = 1...k\}$, where $\mathbf{e}_i \in A^{3k}$ is the ith unit vector of length $3k$. Finally, we transform this generating system into a reduced basis for $\bar{\mathcal{L}}_{3k}$, and then calculate the dimension D for each resolution l.

Example 4.10 **[A Combination of Two or Three Toy Generators]**
In this example, we examine the low-dimensional behavior of the three simple "toy" generators $g_1 = (z, 0, z^3 + z + 1)$, $g_2 = (z^2, 0, z^4 + z + 1)$, and $g_3 = (z^3, 0, z^5 + z^2 + 1)$, as well as the combinations g_{12}, g_{23}, g_{13} of two of them, and the combination g_{123} of all three. Since $\gcd(2^5 - 1,\ 2^4 - 1) = 1$, the periods of g_{12}, g_{23}, and g_{13} are 105, 465, and 217, respectively, and the period of g_{123} is 3255. Table 4.12 gives all values of d, d_i, d_{ij}, and D for $k = 2$. Tables 4.13 and 4.14 have been computed by using the Tables 4.10 and 4.11 and Proposition 4.5. Observe that the values of d, d_i, or d_{ij} never increase when l increases. But this does not necessarily hold for D.

Figure 4.2 shows all the points produced by the generator g_{23}, in dimension $k = 2$. This illustrates the results in the column of g_{23} of Table 4.14. For example, the unit square is equally partitioned into $2^6 = 64$ cells, which corresponds to the equidissection $E_2(3)$. As indicated by Table 4.14, 24 cells contain 8 points, 33 cells contain 7 points, and 7 cells contain 6 points. If we consider the equidissection $E_2(4)$, then, as indicated by Table 4.14, there would be 128 empty cells, while the other cells would contain either 2, 3, or 4 points. □

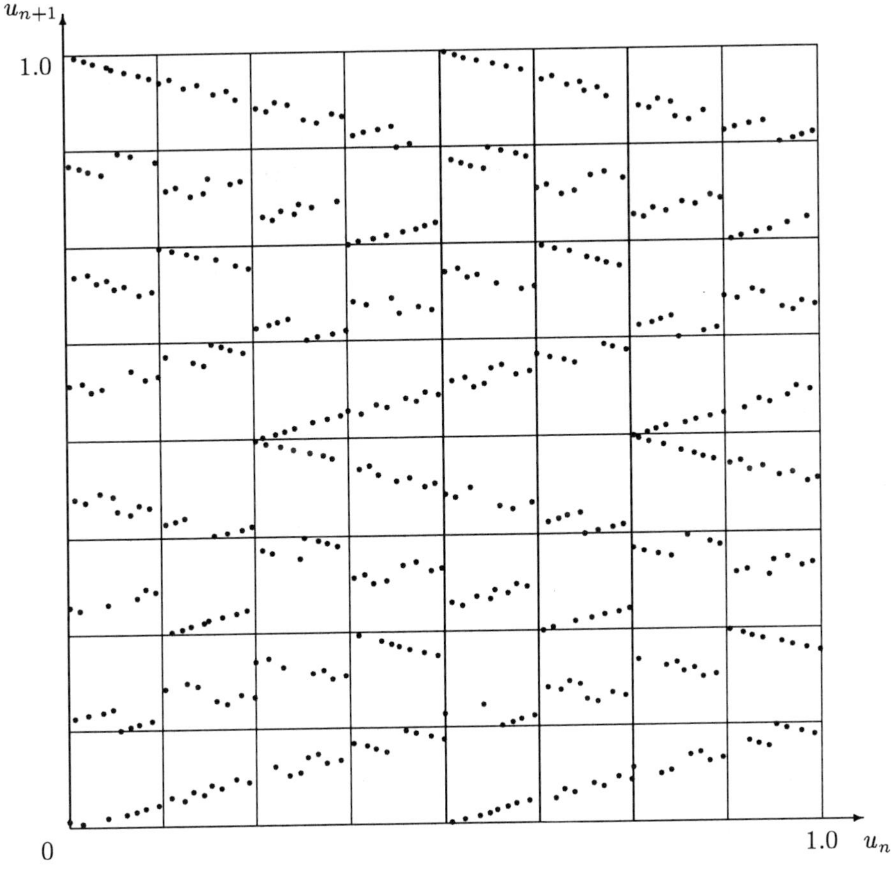

Figure 4.2 The pairs (u_n, u_{n+1}) produced from g_{23}.

Table 4.12 Values of d, d_i, d_{ij}, and D in dimension $k = 2$ for g_1, g_2, g_3, and their combinations

l	d	d_{12}	d_{23}	d_{13}	d_1	d_2	d_3	D
1	10	5	7	6	1	2	3	2
2	8	3	5	4	0	0	1	3
3	6	2	3	2	0	0	0	2
4	4	1	2	1	0	0	0	1
5	2	0	1	0	0	0	0	0
6	0	0	0	0	0	0	0	0

Table 4.13 Nonzero values of $\varphi_{l,k}(m)$ in dimension $k = 2$ for g_1, g_2, and g_3

	g_1		g_2		g_3	
l	m	$\varphi_{l,k}(m)$	m	$\varphi_{l,k}(m)$	m	$\varphi_{l,k}(m)$
1	2	3	4	3	8	3
	1	1	3	1	7	1
2	1	7	1	15	2	15
	0	9	0	1	1	1
3	1	7	1	15	1	31
	0	57	0	49	0	33

Table 4.14 Nonzero values of $\varphi_{l,k}(m)$ in dimension $k = 2$ for g_{12}, g_{23}, g_{13}, and g_{123}

	g_{12}		g_{23}		g_{13}		g_{123}	
l	m	$\varphi_{l,k}(m)$	m	$\varphi_{l,k}(m)$	m	$\varphi_{l,k}(m)$	m	$\varphi_{l,k}(m)$
1	27	1	117	1	55	1	814	3
	26	3	116	3	54	3	813	1
2	7	9	30	1	14	9	204	7
	6	7	29	15	13	7	203	9
3	4	12	8	24	4	28	53	16
	3	17	7	33	3	33	52	16
	2	3	6	7	2	3	51	3
	0	32	0	0	0	0	50	5
	-	-	-	-	-	-	49	20
	-	-	-	-	-	-	48	4
4	2	42	4	84	2	90	16	48
	1	21	3	41	1	37	14	64
	0	65	2	3	0	129	13	4
	-	-	0	128	-	-	12	60
	-	-	-	-	-	-	11	33
	-	-	-	-	-	-	10	35
	-	-	-	-	-	-	9	10
	-	-	-	-	-	-	8	2
5	1	105	2	210	1	217	4	504
	0	919	1	45	0	807	3	246
	-	-	0	769	-	-	2	228
	-	-	-	-	-	-	1	45
	-	-	-	-	-	-	0	1
6	1	105	1	465	1	217	1	3255
	0	3991	0	3631	0	3879	0	841

4.1.3 Resolution-wise Lattice Structure

So far, we have seen that the XOR combination of $LS(2)$ sequences or multiple recursive sequences produces sequences that have good distribution properties in dimensions up to, say 50. In order to analyze the uniform distribution property of such sequences in much higher dimensions, say dimension 1000, even Lenstra's lattice basis reduction algorithm requires a lot of computing power, and is therefore ineffectual with the present computing resource. In this section, we give a solution to this problem, which is based on the new concept of a *resolution-wise lattice*. An important feature of this approach is that it can be used for general GFSR sequences which cannot be analyzed by the conventional 'dimension-wise' lattice approach.[1] In what follows, therefore, we describe our idea with general GFSR sequences. However, we should notice that this new approach can be used only for the analysis of the point set associated with the consecutive k elements, i.e., $(j_1, ..., j_k) = (1, ..., k)$, of GFSR sequences.

We give the matrix representation of general GFSR sequences as follows:

$$\mathbf{x}G, \mathbf{x}CG, ..., \mathbf{x}C^nG, ..., \tag{4.20}$$

where G is an $r \times L$ matrix over $GF(2)$ whose jth column vector, denoted by $G_j, j = 1, ..., L$, is uniquely determined by the equations

$$x_{d_j+i} = \left\langle \mathbf{x}C^{i-1}, {}^tG_j \right\rangle, \text{ for } i = 1, ..., r,$$

equivalent to (4.3). Here $\langle \mathbf{a}, \mathbf{b} \rangle$ means the inner-product of the binary vectors, $\mathbf{a}$ and $\mathbf{b}$, over $GF(2)$ of length r. Following Tootill et al. [87, 101], we give some definitions relevant to the k-dimensional distribution of GFSR sequences. Let $\mathcal{S}(l;k)$ be a set of column vectors $\{C^{i-1}G_j \mid 1 \leq j \leq l (\leq L), 1 \leq i \leq k\}$.

Definition **4.6** A GFSR sequence is said to be k-distributed with l-bit resolution if $\mathcal{S}(l;k)$ is linearly independent but $\mathcal{S}(l+1;k)$ is not. □

Definition **4.7** A GFSR sequence is said to have the order of equidistribution k for the leading l bits if $\mathcal{S}(l;k)$ is linearly independent but $\mathcal{S}(l;k+1)$ is not. □

[1] General GFSRs can be formulated as rth-order MRGs in $GF(2^w)$, as explained in (4.10). Therefore, it might seem that the lattice analysis for MRGs can be applied to general GFSRs. However, this is not the case because the period of GFSRs is $2^r - 1$, not the maximum $2^{rw} - 1$; namely, the lattice points obtained from GFSRs are just a small subset of the entire lattice points of the associated lattices.

First, we give a definition of the resolution-wise lattice of GFSR sequences [92].

DEFINITION **4.8** Let a GFSR sequence be given as in (4.3), and let $g_l(z) = z^{d_l} \pmod{M(z)}$ with $\deg(g_l) < \deg(M)$ for $l = 1, ..., L$. Then we define the resolution-wise lattice of the sequence as follows: for each $l = 1, 2, ..., L$,

$$\mathcal{L}_l = AR_0 + A^l,$$

where $A = GF[2, z]$ and

$$R_0 = (\frac{g_1(z)}{M(z)}, ..., \frac{g_l(z)}{M(z)}).$$

□

The following results show the usefulness of this notion.

Proposition 4.6 *For a GFSR sequence,* $u_n, n = 1, 2, ...,$ *in* (4.3), *define* $R_n = zR_{n-1} \pmod 1), n = 1, 2, ...,$ *where the modulo operation is applied component-wise. Let* $\varphi_{l,k}(m)$ *correspond to the point set* $S = \{(u_n, ..., u_{n+k-1}), n = 1, 2, ...\}$, *and let* $\bar{\varphi}_{k,l}(m)$ *correspond to the point set* $\bar{S} = \{R_n = (s_n^{(1)}, ..., s_n^{(l)}), n = 1, 2, ...,\}$. *Then we have*

$$\varphi_{l,k}(m) = \bar{\varphi}_{k,l}(m).$$

Proof: This follows from the fact that both $\varphi_{l,k}(m)$ and $\bar{\varphi}_{k,l}(m)$ correspond to $\mathcal{S}(l; k)$.□

Since the evolution of $R_n, n = 1, 2, ...,$ gives $2^r - 1$ distinct points in the l-dimensional resolution-wise lattice, we obtain the following result:

Proposition 4.7 *Let* k_l *be the last successive minimum of the* l*-dimensional resolution-wise lattice associated with a GFSR sequence. Then the sequence has the order of equidistribution* $-k_l$ *for the leading* l *bits,*

Since $LS(2)$ sequences can be formulated as GFSR sequences, the k-dimensional distribution of this class of sequences can be analyzed by using the dimension-wise lattice structure as well as the resolution-wise one. Taking this case, we consider the advantage of the resolution-wise lattice approach in the light of practical efficiency. Lenstra [47] showed that his basis reduction algorithm runs in $O(B^2k^4)$, where B is the degree of $M(z)$ and k is the maximum number of dimensions. Therefore, in the case of the conventional 'dimension-wise' lattice analysis for the high-dimensional behavior of long-period $LS(2)$ sequences, the values of k and B become usually 500 or more. On the other hand, in the case of resolution-wise lattice analysis, while B is just as large, k is at most 32 (for a 32-bit computer). Obviously, the latter approach is practically much more efficient.

We have the following results concerning the relation between resolution-wise and dimension-wise lattices for any $LS(2)$ sequence.

Proposition 4.8 *Let $k_1(l)$ and $k_l(l)$ be the first and last successive minima for the resolution-wise lattice in resolution l associated with a $LS(2)$ sequence. If $k_1(l) > k_1(l-1)$, then the dimension-wise lattice associated with the sequence has the first successive minimum equal to $-l$ in dimension k for $|k_1(l-1)| > k \geq |k_1(l)|$. If $k_{l-1}(l-1) < k_l(l)$, then the dimension-wise lattice has the last successive minimum equal to $-(l-1)$ in dimension k for $|k_{l-1}(l-1)| \geq k > |k_l(l)|$; in other words, the sequence is k-distributed with $-(l-1)$-bit resolution for $|k_{l-1}(l-1)| \geq k > |k_l(l)|$.*

Proposition 4.9 *Let $l_1(k)$ and $l_k(k)$ be the first and last successive minima for the dimension-wise lattice in dimension k associated with a $LS(2)$ sequence. If $l_1(k) > l_1(k-1)$, then the resolution-wise lattice associated with the sequence has the first successive minimum equal to $-k$ in resolution l for $|l_1(k-1)| > l \geq |l_1(k)|$. If $l_{k-1}(k-1) < l_k(k)$, then the resolution-wise lattice has the last successive minimum equal to $-(k-1)$ in resolution l for $|l_{k-1}(k-1)| \geq l > |l_k(k)|$; in other words, the equidistribution order of the sequence is $-(k-1)$ in resolution l for $|l_{k-1}(k-1)| \geq l > |l_k(k)|$.*

On the basis of these two results, we can derive l_1 and l_k of the dimension-wise lattice from k_1 and k_l of the resolution-wise lattice, and vice versa.

EXAMPLE **4.11 [Four GFSR Generators with $r = 521$]**
First, we apply the resolution-wise lattice analysis to two Tausworthe generators with $r = 521$. Tootill et al. [100] recommended these generators because they produce sequences that are uniformly distributed in dimension $m = [r/s]$, and in addition, their subsequences are also uniformly distributed in dimension

m. The respective specifications of the generators, Tootill I and II, are $s = 32$ and 48 with $d = s$. It is easily confirmed that the GFSR implementation with trinomial $(z^{521} + z^{32} + 1)$ and pentanomial $(z^{521} + z^{48} + z^{32} + z^{16} + 1)$ recurrence relations can be harnessed for Tootill I and II, respectively.

Second, we analyze the Arvillias-Maritsas generator [7] and its variant by Fushimi-Tezuka [30]. The former is defined as a Lewis-Payne sequence,

$$u_n = .x_{d+n}x_{2d+n}\cdots x_{32d+n},$$

where $d \approx \frac{9}{32} \times 2^{521}$. Since the period $2^{521} - 1$ is a Mersenne prime, this sequence can be formulated as a Tausworthe sequence. Figure 4.3 illustrates the hardware implementation of this generator. The specification is as follows:

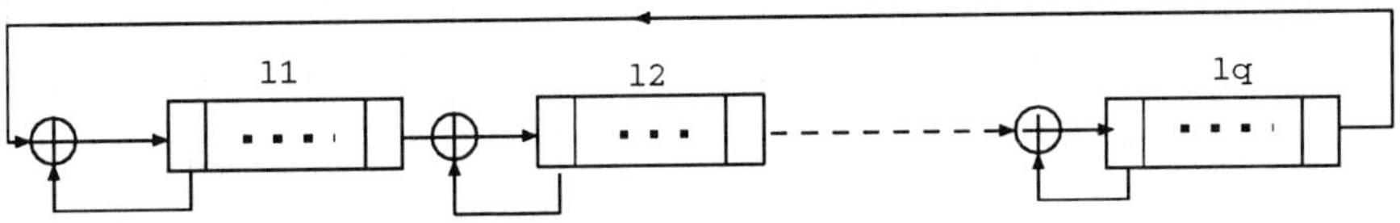

Figure 4.3 Toggle register for the Arvillias-Maritsas generator.

$z^{521} + z^{32} + 1$, and $l_1 = l_2 = \cdots = l_{31} = 1$ and $l_{32} = 490$, which was called a "realistic generator" in Arvillias and Maritsas. The reason for the name is the fact that the autocorrelation of the sequence is very small up to the delay of $(2^{521} - 1)/32$, which is the maximum possible. Figure 4.4 shows the very bad uniform distribution of the consecutive pairs $(u_n, u_{n+1}), n = 1, 2, ..., 2000$ from this sequence. Here we should notice that the autocorrelation with delay 1 of the sequence is very small because of the symmetry of the point distribution with respect to the line $u_n = \frac{1}{2}$. The improved variant by Fushimi and Tezuka is given as $l_i = 17$ for $i = 4, 8, 11, 15, 18, 22, 25, 29, 32$, and $l_i = 16$ for other i. Thus

$$u_n = .x_{d_1+n}x_{d_2+n}\cdots x_{d_{32}+n},$$

where $d_i = (i-1)d + \sum_{j=1}^{i-1} l_i$ for $i = 1, 2, ..., 32$, and d is a multiple of $2^{521}/32$, which is less than $2^{521} - 1$. Since $d_2 - d_1 = d + 16$, but $d_5 - d_4 = d + 17$, the sequence cannot be formulated as a Tausworthe sequence or as a Lewis-Payne sequence. However, it can be formulated as a GFSR sequence.

Table 4.15 shows the first and last successive minima k_1 and k_l of the resolution-wise lattice in resolutions of 2 to 32 bits for these four generators. All of them are unsatisfactory in almost all resolutions.

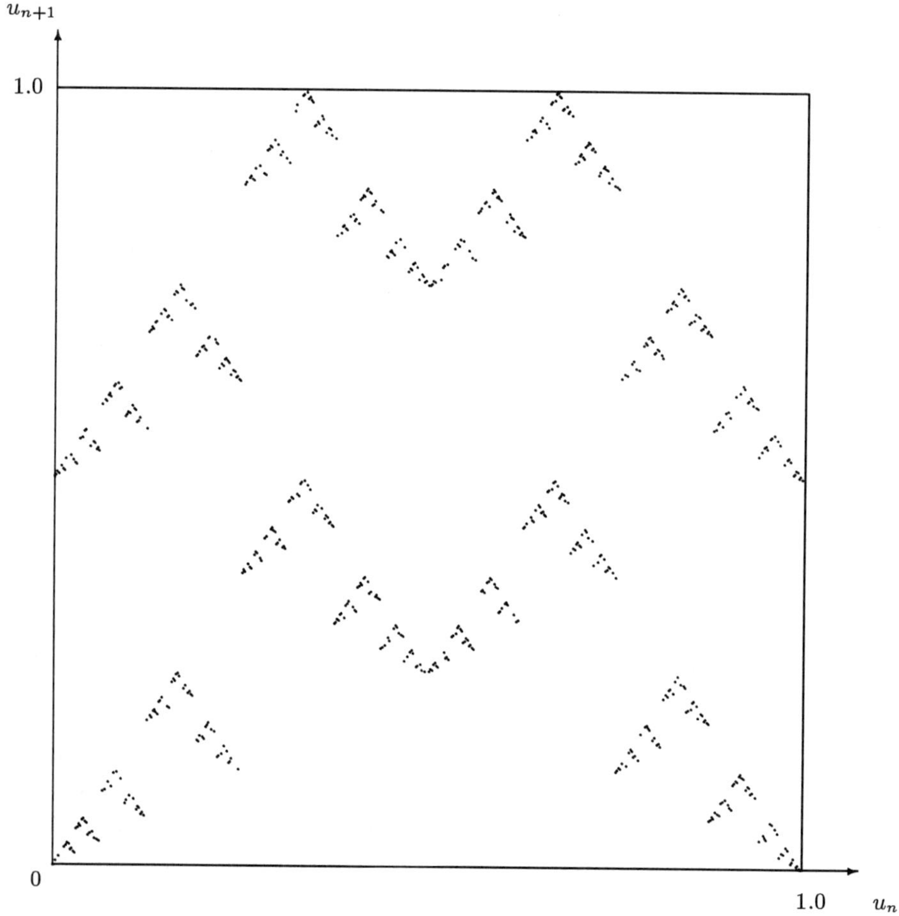

Figure 4.4 A set of two-dimensional points, $(u_n, u_{n+1}), n = 1, 2, ..., 2000$, from Arvillias and Maritsas's "realistic generator."

Table 4.15 Values of k_1 and k_l in resolutions $l = 2$ to 31 for Tootill I, II, Arvillias-Maritsas, and Fushimi-Tezuka generators

Res. l		2	3	4	5	6	7	8	9	10	11
$\lfloor 521/l \rfloor$		260	173	130	104	86	74	65	57	52	47
Tootill	$-k_l$	114	114	65	56	49	49	49	49	16	16
I	$-k_1$	407	293	179	114	114	114	114	65	65	65
Tootill	$-k_l$	76	76	76	76	76	65	11	11	11	11
II	$-k_1$	445	369	293	217	141	76	76	76	76	76
Arvillias	$-k_l$	1	1	1	1	1	1	1	1	1	1
-Maritsas	$-k_1$	520	519	518	517	516	515	514	513	512	511
Fushimi	$-k_l$	16	16	16	16	16	16	16	16	16	16
-Tezuka	$-k_1$	505	489	473	456	440	424	408	391	375	359

Res. l		12	13	14	15	16	17	18	19	20	21
$\lfloor 521/l \rfloor$		43	40	37	34	32	30	28	27	26	24
Tootill	$-k_l$	16	16	16	16	16	16	16	16	16	16
I	$-k_1$	65	65	49	49	49	49	49	49	49	49
Tootill	$-k_l$	11	11	11	11	11	11	11	11	11	11
II	$-k_1$	76	65	65	65	65	65	65	65	54	54
Arvillias	$-k_l$	1	1	1	1	1	1	1	1	1	1
-Maritsas	$-k_1$	510	509	508	507	506	505	504	503	502	501
Fushimi	$-k_l$	16	16	16	16	16	16	16	16	16	16
-Tezuka	$-k_1$	342	326	310	294	277	261	245	228	212	196

Res. l		22	23	24	25	26	27	28	29	30	31
$\lfloor 521/l \rfloor$		23	22	21	20	20	19	19	17	17	16
Tootill	$-k_l$	16	16	16	16	16	16	16	16	16	16
I	$-k_1$	49	33	33	33	33	33	33	33	33	33
Tootill	$-k_l$	11	11	11	11	11	11	11	11	11	11
II	$-k_1$	54	54	54	54	54	43	43	43	43	43
Arvillias	$-k_l$	1	1	1	1	1	1	1	1	1	1
-Maritsas	$-k_1$	500	499	498	497	496	495	494	493	492	491
Fushimi	$-k_l$	16	16	16	16	16	16	16	16	16	16
-Tezuka	$-k_1$	180	163	147	131	114	98	82	66	49	33

□

EXAMPLE **4.12 [Two Twisted GFSR Generators]**
Twisted GFSR generators were recently introduced by Matsumoto and Kurita [60]:

$$\mathbf{v}_n = \mathbf{v}_{n-r+s} \text{ XOR } \mathbf{v}_{n-r}A, \text{ for } n = r+1, r+2, ..., \tag{4.21}$$

where $\mathbf{v}_n, n = 1, 2, ...,$ is a sequence of vectors in $GF(2^w)$, A is a $w \times w$ matrix over $GF(2)$, and $r > s$. In particular, they analyzed the following special case, which is very useful for quick generation of the sequence:

$$\mathbf{v}_n = \mathbf{v}_{n-r+s} \text{ XOR } \mathbf{v}_{n-r}{}^tC, \tag{4.22}$$

where C is a $w \times w$ companion matrix, whose wth column vector is $\mathbf{a} = (a_0, ..., a_{w-1})$. The conversion from a binary vector $\mathbf{v} = (v_0, ..., v_{w-1})$ into a random number u in $[0, 1)$ is as follows:

$$u = \frac{1}{2^w} \sum_{i=0}^{w-1} v_i 2^{w-1-i}.$$

The parameters r, s, and C are chosen so that the period of the sequence $\mathbf{v}_n, n = 1, 2, ...,$ becomes the maximum of $2^{rw} - 1$. The implementation is described as follows: Let $\mathbf{v}_n = (v_{n,0}, ..., v_{n,w-1})$. The rest is very simple:

$$\begin{aligned} \text{if } v_{n-r,w-1} = 0 \text{ then } \mathbf{v}_n &= \mathbf{v}_{n-r+s} \text{ XOR } (\mathbf{v}_{n-r} >> 1) \\ \text{else } \mathbf{v}_n &= \mathbf{v}_{n-r+s} \text{ XOR } (\mathbf{v}_{n-r} >> 1) \text{ XOR } \mathbf{a}. \end{aligned}$$

One advantage of Twisted GFSRs is that they use the minimal number r of w-bit words which is necessary to produce the period length $2^{rw} - 1$. In this sense, the scheme can be viewed as an improved version of trinomial-type GFSR sequences because the latter produces a period of $2^r - 1$ with an identically sized array of r w-bit words. In addition, the generation speed is almost the same.

We analyze the uniform distribution of the following two generators, Twisted GFSR I and II, with the maximum periods: one is $(w, r, s) = (31, 13, 2)$ with $(0, a_0, ..., a_{30}) =$ 6B5ECCF6 (in hexadecimal), and the other is $(w, r, s) = (32, 25, 7)$ with $(a_0, ..., a_{31}) =$ 8EBFD028 (in hexadecimal). The respective periods are $2^{403} - 1$ and $2^{800} - 1$.

□

Table 4.16 Values of k_1 and k_l in resolutions $l = 2$ to 31 for the Twisted GFSR with a period of $2^{403} - 1$

Res. l		2	3	4	5	6	7	8	9	10	11
$\lceil 403/l \rceil$		201	134	100	80	67	57	50	44	40	36
Twisted	k_l	13	13	13	13	13	13	13	13	13	13
GFSR I	k_1	390	377	364	351	338	325	312	299	286	273

Res. l		12	13	14	15	16	17	18	19	20	21
$\lceil 403/l \rceil$		33	31	28	26	25	23	22	21	20	19
Twisted	k_l	13	13	13	13	13	13	13	13	13	13
GFSR I	k_1	260	247	234	221	208	195	182	169	156	143

Res. l		22	23	24	25	26	27	28	29	30	31
$\lceil 403/l \rceil$		18	17	16	16	15	14	14	13	13	13
Twisted	k_l	13	13	13	13	13	13	13	13	13	13
GFSR I	k_1	130	117	104	91	78	65	52	39	26	13

Table 4.17 Values of k_1 and k_l in resolutions $l = 2$ to 31 for the Twisted GFSR with a period of $2^{800} - 1$

Res. l		2	3	4	5	6	7	8	9	10	11
$\lceil 800/l \rceil$		400	266	200	160	133	114	100	88	80	72
Twisted	k_l	25	25	25	25	25	25	25	25	25	25
GFSR II	k_1	775	750	725	700	675	650	625	600	575	550

Res. l		12	13	14	15	16	17	18	19	20	21
$\lceil 800/l \rceil$		66	61	57	53	50	47	44	42	40	38
Twisted	k_l	25	25	25	25	25	25	25	25	25	25
GFSR II	k_1	525	500	475	450	425	400	375	350	325	300

Res. l		22	23	24	25	26	27	28	29	30	31
$\lceil 800/l \rceil$		36	34	33	32	30	29	28	27	26	25
Twisted	k_l	25	25	25	25	25	25	25	25	25	25
GFSR II	k_1	275	250	225	200	175	150	125	100	75	50

As is clear from Tables 4.16 and 4.17, both of the two generators perform badly in dimensions higher than r. In fact, it is easy to derive the following general result [93, 99]:

Proposition 4.10 *Any Twisted GFSR in* (4.22) *is* k*-distributed with at most 2-bit resolution for all* $k > r$.

This result implies the following important consequence: Twisted GFSR sequences can be viewed as GFSR sequences with period of $2^{rw} - 1$ whose characteristic polynomials have many nonzero terms. In fact, the generator used in Table 4.17 is formulated as a GFSR whose characteristic polynomial has 93 nonzero terms. The above result indicates that even GFSR sequences with many-term recurrence relations do not necessarily produce good sequences with respect to uniform distribution. Furthermore, Tezuka and Wong [99] showed experimentally that Twisted GFSRs suffer from the same problem as AWC-SWB generators for Ising-model Monte Carlo simulations with the Wolff algorithm.

Combined GFSR Sequences

In the foregoing section, we saw that trinomial-type GFSR sequences as well as Twisted GFSR sequences are unsatisfactory in their high-dimensional distribution. The main reason for this seems the trade-off between two (somewhat contradictory) requirements: fast generation and good uniform distribution in high dimensions. For example, GFSR sequences with trinomial recurrence relations can be very quickly generated, but their distribution properties in high dimensions are miserable, as we have seen. This is also true for Twisted GFSR sequences. Therefore, we consider the XOR combination of GFSR sequences at the price of speed, since the same approach was previously successful for $LS(2)$ sequences. In general, we define the combined GFSR sequence as follows: for $n = 1, 2, ...,$

$$U_n = u_n^{(1)} \text{ XOR } \cdots \text{ XOR } u_n^{(J)},$$

where for $j = 1, ..., J$, each sequence $u_n^{(j)}, n = 1, 2, ...,$ is a GFSR sequence with a period of $2^{r_j} - 1$. Here we assume that the periods of components are pairwise coprime, so that the period of the combined sequence $U_n, n = 1, 2, ...,$ is equal to the product of all the periods of $u_n^{(j)}, j = 1, ..., J$.

For such a combined generator, we need the analysis method of its uniform distribution. We are now in a position to give the main result:

Proposition 4.11 *For a combined GFSR sequence, $U_n, n = 1, 2, ...,$ define $R_n = R_n^{(1)} \otimes \cdots \otimes R_n^{(J)}$, where each $R_n^{(j)}$ corresponds to a component GFSR sequence $u_n^{(j)}, n = 1, 2, ...,$ for $j = 1, ..., J$, and $\otimes$ means the coordinate-wise application of* XOR . *Let $\varphi_{l,k}(m)$ correspond to the point set $S = \{(U_n, ..., U_{n+k-1}),$ $n = 1, 2, ...\}$, and let $\bar{\varphi}_{k,l}(m)$ correspond to the point set $\bar{S} = \{R_n = (s_n^{(1)}, ..., s_n^{(l)}),$ $n = 1, 2, ...,\}$. Then we have*

$$\varphi_{l,k}(m) = \bar{\varphi}_{k,l}(m).$$

Proof: This follows from the fact that we have, modulo one,

$$\begin{aligned} R_n &= R_n^{(1)} \otimes \cdots \otimes R_n^{(J)} \\ &= zR_{n-1}^{(1)} \otimes \cdots \otimes zR_{n-1}^{(J)} \\ &= z(R_{n-1}^{(1)} \otimes \cdots \otimes R_{n-1}^{(J)}) \\ &= zR_{n-1}, \end{aligned}$$

where the modulo operation is applied component-wise. □

Let k_l be the last successive minimum of the l-dimensional resolution-wise lattice associated with the combined GFSR sequence. Let $k_1^{(-j)}$ be the first successive minimum of the l-dimensional resolution-wise lattice associated with the combined sequence of all $u_n^{(i)}, n = 1, 2, ...$ for $1 \leq i \neq j \leq J$. Let $r = r_1 + \cdots + r_J$. The following theorem is the counterpart of Theorem 4.3 for combined $LS(2)$ sequences;

Theorem 4.4 *If $\min_{j=1}^{J} k_1^{(-j)} \geq k_l$, then for k with $\min_{j=1}^{J} k_1^{(-j)} \geq k \geq k_l$, the number of points $(U_n, ..., U_{n+k-1}), n = 1, 2, ...,$ that are contained in each cell of the equidissection $E_{|k|}(l)$ is equal to an integer in the range $[2^{r+lk} - J, 2^{r+lk}]$.*

EXAMPLE **4.13 [continued from Example 4.8]**
This is a continuation of Example 4.8, where we performed the dimension-wise lattice analysis in dimensions up to 40 for the combined MRG. It consumed a lot of CPU time on an RS/6000. However, by using almost the same amount of CPU time, we can apply the resolution-wise lattice analysis to the same combined MRG for much higher dimensions. The results are given in Table 4.18, which shows that the combined MRG has very good uniform distributions in dimensions higher than 40 as well as lower than 40 already shown before.

□

Table 4.18 Values of k_1 and k_l in resolutions $l = 2$ to 29 for the combined and component MRGs

	Res. l	2	3	4	5	6	7	8	9	10	11	12	13	14	15
	$\lfloor 358/l \rfloor$	179	119	89	71	59	51	44	39	35	32	29	27	25	23
M	$-k_l$	178	119	89	71	59	51	44	39	35	32	29	27	25	23
	$-k_1$	180	120	91	72	60	52	46	40	36	33	30	28	26	24
	$\lfloor 155/l \rfloor$	77	51	38	31	25	22	19	17	15	14	12	11	11	10
M_1	$-k_l$	77	51	38	30	25	22	19	14	14	12	12	10	5	5
	$-k_1$	78	52	40	32	26	23	20	19	16	15	14	13	13	13
	$\lfloor 203/l \rfloor$	101	67	50	40	33	29	25	22	20	18	16	15	14	13
M_2	$-k_l$	101	67	50	40	33	28	25	21	19	13	13	12	7	7
	$-k_1$	102	69	52	41	35	30	26	24	22	20	19	19	19	19

	Res. l	16	17	18	19	20	21	22	23	24	25	26	27	28	29
	$\lfloor 358/l \rfloor$	22	21	19	18	17	17	16	15	14	14	13	13	12	12
M	$-k_l$	22	21	19	18	17	16	16	15	14	14	13	12	12	12
	$-k_1$	23	22	20	19	18	18	17	16	15	15	14	14	13	13
	$\lfloor 155/l \rfloor$	9	9	8	8	7	7	7	6	6	6	5	5	5	5
M_1	$-k_l$	5	5	5	5	5	5	5	5	5	5	5	5	5	5
	$-k_1$	13	13	12	12	12	12	12	9	9	9	8	8	8	8
	$\lfloor 203/l \rfloor$	12	11	11	10	10	9	9	8	8	8	7	7	7	7
M_2	$-k_l$	7	7	7	7	7	7	7	7	7	7	7	7	7	7
	$-k_1$	19	19	19	14	13	13	12	12	12	12	12	12	12	7

4.1.4 Further Analysis via Dyadic Boxes

So far, we have analyzed the uniform distribution of $LS(2)$ sequences as well as GFSR sequences by means of the notion of equidissection. In this section, we will introduce a much deeper analysis of uniform distribution based on dyadic boxes, which are defined as, for nonnegative integers $e_1, ..., e_k$,

$$J_k(e_1, ..., e_k) = \left\{ \prod_{h=1}^{k} [i_h 2^{-e_h}, (i_h + 1)2^{-e_h}) \mid 0 \le i_h \le 2^{e_h} - 1, 1 \le h \le k \right\}.$$

Note that the equidissection $E_k(l)$ is a special case of $J_k(e_1, ..., e_k)$; that is, $E_k(l) = J_k(l, ..., l)$. We define a figure of merit $\rho^{(k)}$ for dyadic box analysis as follows:

DEFINITION **4.9** *Denote a GFSR sequence with a period* $2^r - 1$ *by* $u_n, n = 1, 2, ..., 2^r - 1$. *For the point set* $P = \{(u_n, ..., u_{n+k-1}), n = 1, ..., 2^r - 1\} \cup \{0\}$, *we define the figure of merit* $\rho^{(k)}$ *as the maximum of* t *such that each cell of* $J_k(e_1, ..., e_k)$ *contains the same number of points for all integers* $0 \le e_1, ..., e_k \le t$ *with* $\sum_{j=1}^{k} e_j = t$.

This means that every cell of $J_k(e_1, ..., e_k)$ with the volume of 2^{-t} contains exactly 2^{r-t} points of P for all $0 \le t \le \rho^{(k)}$. Hence, the larger $\rho^{(k)}$, the more uniform the sequences. The following is an extension of $\mathcal{S}(l; k)$:

DEFINITION **4.10** *Recall C^{i-1} and G_j. Define*

$$\mathcal{S}(e_1, ..., e_k) = \{C^{i-1}G_j \mid 0 \le j \le e_i, 1 \le i \le k\},$$

where $C^{i-1}G_0$ is an empty element.

We have the following result [72, 86, 88]:

Proposition 4.12 *The figure of merit $\rho^{(k)}$ is equal to the maximum of t such that the rank of $\mathcal{S}(e_1, ..., e_k)$ is full for all k-tuples of integers $(e_1, ..., e_k)$ with $\sum_{i=1}^k e_i = t$ and $0 \le e_1, ..., e_k \le t$.*

For $LS(2)$ sequences, the above result can be rewritten in terms of lattice parameters. Here we define $\deg(0) = -1$.

Corollary 4.3 *We have*

$$\rho^{(k)} = k - 1 + \min \sum_{i=1}^{k} \deg(h_i),$$

where the minimum is taken over all nonzero solutions $(h_1(z), ..., h_k(z))$ of the equation:

$$\sum_{i=1}^{k} h_i(z) g(z)^{i-1} = 0 \pmod{M(z)}.$$

Proof: Let $W = g(C)$, where C is the companion matrix whose characteristic polynomial is $M(z)$. Then the matrix representation for the leading r-bits of $LS(2)$ sequences is given by

$$\mathbf{x}, \mathbf{x}W, ..., \mathbf{x}W^n, ...$$

Let $\mathbf{h}_i$ be a binary vector $(h_{i1}, ..., h_{im})$, and define a polynomial $h_i(z) = \sum_{j=1}^m h_{ij} z^{j-1}$. Then, we find that there exists a nonzero solution $(\mathbf{h}_1, ..., \mathbf{h}_k)$ for

$$\mathbf{h}_1 {}^tW + \cdots + \mathbf{h}_k ({}^tW)^k = 0$$

if and only if there exist a nonzero solution $(h_1(z), ..., h_k(z))$ for

$$h_1(z)g(z) + \cdots + h_k(z)g(z)^k = 0 \pmod{M(z)}. \tag{4.23}$$

Suppose that $(e_1, ..., e_k)$ is such that $\sum_{i=1}^{k} e_j = l$ is minimum taken over all $0 \le e_1, ..., e_k \le l$ satisfying that the rank $\mathcal{S}(e_1, ..., e_k)$ is not full. Then, there exists a nonzero solution $(h_1(z), ..., h_k(z))$ with $\deg(h_i) = e_i - 1$ for (4.23). Hence the proof is complete.□

The point set obtained from an $LS(2)$ sequence consists of lattice points, as explained in the preceding sections. The associated lattice $\mathcal{L}_k$ is spanned by the basis of k vectors, $\mathbf{e}_1 = (1/M(z),\ g(z)/M(z), ..., g(z)^{k-1}/M(z))$, $\mathbf{e}_2 = (0, 1, 0, ..., 0)$, ..., $\mathbf{e}_k = (0, ..., 0, 1)$. Namely, the lattice point is given by

$$\sum_{i=1}^{k} b_i(z)\mathbf{e}_i,$$

where $b_i(z), i = 1, ..., k$, are polynomials over $GF(2)$. The dual lattice $\mathcal{L}_k^*$ is spanned by k vectors $\mathbf{e}_1^* = (M(z), 0, ..., 0)$, $\mathbf{e}_2^* = (g(z), 1, 0, ..., 0)$, ..., $\mathbf{e}_k^* = (g(z)^{k-1}, 0, ..., 0, 1)$. The lattice point is given by

$$\sum_{i=1}^{k} b_i(z)\mathbf{e}_i^*,$$

where $b_i(z), i = 1, ..., k$, are polynomials over $GF(2)$. Thus, Corollary 4.3 says that we have

$$\rho^{(k)} = |s(\mathcal{L}_k^*)|_{L1} + k - 1,$$

where $|s(\mathcal{L}_k^*)|_{L1}$ denotes the length of the nonzero shortest vector with respect to the L_1-norm of the dual lattice $\mathcal{L}_k^*$. Therefore, we see that the figure of merit $\rho^{(k)}$ for $LS(2)$ sequences can be viewed as the counterpart of $\tilde{\nu}_k$, which is a measure of the lattice structure of linear congruential sequences discussed in Chapter 3.

We will show an interesting relation between the figure of merit and the discrepancy of GFSR sequences [72, 86], which can be derived through the more general result for a (t, m, k)-net in base b, to be developed in Chapter 6.

Theorem 4.5 *The k-dimensional discrepancy of GFSR sequences over the entire period $N = 2^r - 1$ is given as*

$$D_N^{(k)} \le c(k)\frac{(\log N)^{k-1}}{2^{\rho^{(k)}}},$$

where $c(k)$ *is a constant depending only on* k.

Hence, the value of $\rho^{(k)}$ can be regarded as a figure of merit for the discrepancy of the sequences; that is to say, the larger $\rho^{(k)}$ is, the lower the discrepancy. According to Niederreiter's net theory [72], we have a more explicit relation for the two-dimensional case.

$$D_N^{(2)} \leq [\frac{\rho^{(2)} + 3}{2}] \frac{1}{2^{\rho^{(2)}}},$$

where $N = 2^r$, that is, we assume that the origin is added to the point set, and $r = \deg(M)$.

The problem encountered when we consider practical applications of the dyadic box analysis to $LS(2)$ sequences (as well as GFSR sequences) lies in the inefficiency of computing the figure of merit $\rho^{(k)}$, in other words, of finding a shortest nonzero vector with respect to the L_1-norm (the sum of the degrees of all components) of a polynomial lattice. Although the problem of finding a shortest vector with respect to the L_∞ norm can be efficiently solved by using the Lenstra algorithm given in Chapter 2, at present no polynomial time algorithms are known for the case of the L_1-norm. The only exception is the two-dimensional case, for which the next theorem, see [95], links the continued fraction expansion of $g(z)/M(z)$ with the two-dimensional figure of merit.

Theorem 4.6 *Consider an* $LS(2)$ *sequence from* $G = (g, 0, M)$. *Let the partial quotients in the continued fraction expansion over* $GF(2)$ *of* $g(z)/M(z)$ *be* $A_1(z)$, ..., $A_s(z)$, *that is,*

$$\frac{g(z)}{M(z)} = \cfrac{1}{A_1(z) + \cfrac{1}{A_2(z) + \ddots + \cfrac{1}{A_s(z)}}}.$$

Then we have

$$\rho^{(2)} = r + 1 - \max_{1 \leq i \leq s} \deg(A_i),$$

where r *is the degree of* $M(z)$.

There exists a pair of polynomials $(g(z), M(z))$ for any degree of $M(z)$ such that the degrees of partial quotients in the continued fraction expansion of $g(z)/M(z)$ are all 1. In this context, Fibonacci polynomials over $GF(2)$ were defined as follows [95]:

Definition **4.11** A pair of polynomials $(a(z), b(z))$ with $\deg(a) < \deg(b)$ is called a pair of Fibonacci polynomials over $GF(2)$ if the partial quotients in the continued fraction expansion over $GF(2)$ of $a(z)/b(z)$ are all of degree 1. □

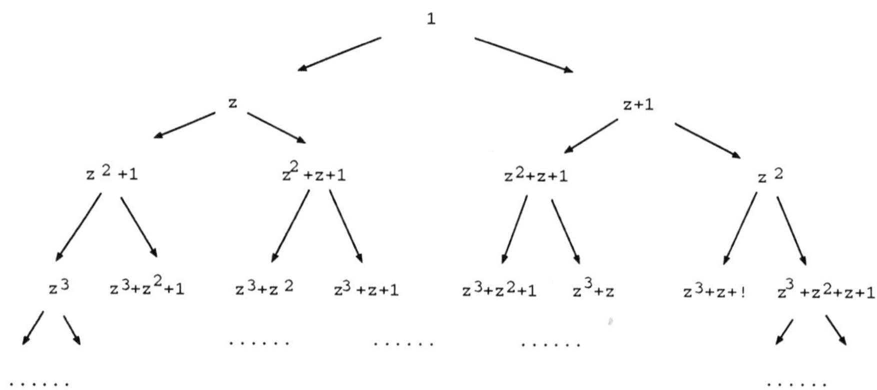

Figure 4.5 Tree of Fibonacci polynomials.

The following recurrence relation produces a sequence of Fibonacci polynomials, $F_i(z), i = 0, 1, 2, \ldots$.

$$F_i(z) = A_i(z)F_{i-1}(z) + F_{i-2}(z), \quad i = 2, 3, \ldots, \tag{4.24}$$

where $F_0(z) = 1, F_1(z) = A_1(z)$, and $A_i(z), i = 1, 2, \ldots$, are arbitrary polynomials over $GF(2)$ of degree 1, that is, $A_i(z) = z$ or $z + 1$. Thus, $(F_i(z), F_{i+1}(z))$ is a pair of Fibonacci polynomials. Figure 4.5 shows the initial part of a tree of Fibonacci polynomials.

In previous papers [94, 95], we considered two cases where $g(z)$ satisfies the trinomial relation

$$g(z)^r + g(z)^s + 1 = 0 \pmod{M(z)}, \tag{4.25}$$

with $r = \deg(M) > s > 0$, or the pentanomial relation

$$g(z)^r + g(z)^{s_1} + g(z)^{s_2} + g(z)^{s_3} + 1 = 0 \pmod{M(z)}, \tag{4.26}$$

with $r = \deg(M) > s_1 > s_2 > s_3 > 0$. For the former, we have the generation scheme:

$$u_{n+r} = u_{n+s} \text{ XOR } u_n.$$

For the latter, we have the generation scheme

$$u_{n+r} = u_{n+s_1} \text{ XOR } u_{n+s_2} \text{ XOR } u_{n+s_3} \text{ XOR } u_n.$$

Both cases provide us with an efficient generation scheme of $LS(2)$ sequences whose parameters are pairs of Fibonacci polynomials. Tables 4.19 and 4.20 summarize some good parameters.

Now, we explain how to find these parameters. Mesirov and Sweet [61] proved that exactly two $g(z)$ exist for every irreducible $M(z)$, where $(g(z), M(z))$ is a pair of Fibonacci polynomials. Therefore, we search for a pair $(g(z), M(z))$ satisfying the condition (4.25) or (4.26). The searching strategy is as follows: by using the recurrence (4.24), we generate all pairs $(F_{r-1}(z), F_r(z))$, and then for each pair, we check whether

$$F_{r-1}(z)^r + F_{r-1}(z)^s + 1 = 0 \pmod{F_r(z)},$$

which come from the condition (4.25). Or, for the pentanomial case, we check whether

$$F_{r-1}(z)^r + F_{r-1}(z)^{s_1} + F_{r-1}(z)^{s_2} + F_{r-1}(z)^{s_3} + 1 = 0 \pmod{F_r(z)},$$

which come from the condition (4.26). If the check is OK, then the obtained pair was regarded as $(g(z), M(z))$ for the $LS(2)$ sequence.

For each degree r with $5 \le r \le 21$, we found several pairs of Fibonacci polynomials satisfying either (4.25) or (4.26) that pass the above check. Then, we selected the best pair by using the following procedure: the minimum t such that $\rho^{(k)} \ge r - k - t$ for all $3 \le k \le 5$, is calculated for each of the obtained generators, and then the generator giving the smallest t is selected for each degree. If more than one generator exists after that selection, choose the generator with the best figures of merit in dimensions 5 and 6. By this procedure, we selected exactly one generator for each degree $5 \le r \le 21$. The corresponding polynomials $M(z)$ and $g(z)$ are given in Table 4.19, where we list only the exponents of the nonzero terms of the polynomials. In this search, we omitted a pair $(g'(z), M'(z))$, which is obtained by the transformation $(g'(z), M'(z)) = (g(z+1), M(z+1))$. We also omitted the pair for the reciprocal case, $z^r + z^{r-s} + 1$, or $z^r + z^{r-s_1} + z^{r-s_2} + z^{r-s_3} + 1$, which produces the sequence u_n in reverse order. In the range $22 \le r \le 32$, since the computational time grows too large, we restricted ourselves to the trinomial case (4.25), and found that there exist pairs of Fibonacci polynomials that pass the above check only for $r = 22, 23, 25, 28$, and 31. For each degree, the best pair is selected by using the same criterion as above. The corresponding polynomials $M(z)$ and $g(z)$ are given in Table 4.19.

In Table 4.20, we summarize the figures of merit in dimensions 2 through 6 for all these generators. Since no efficient algorithm is currently available for calculation of the figures of merit in dimensions higher than two, we used the brute force calculation, which is the same as Andre et al.'s algorithm [5]. This may be described as follows:

> Step 1: Set $t = r$.
> Step 2: If the system $\mathcal{S}(e_1, ..., e_k)$ is linearly independent for all k-tuples $(e_1, ..., e_k)$ satisfying $\sum_{i=1}^{k} e_i = t$ and $0 \leq e_1, ..., e_k \leq t$, then output $\rho^{(k)} = t$ (and stop).
> Step 3: $t = t - 1$, and goto Step 2.

Table 4.19 Pairs of Fibonacci polynomials ($g(z)$, $M(z)$) for $LS(2)$ sequences with $5 \leq r \leq 31$

$G_{5,3,2,1}$	M : 0 3 5
	g : 1 4
$G_{6,5,2,1}$	M : 0 1 6
	g : 0 1 2 3 5
$G_{7,5,3,1}$	M : 0 1 2 5 7
	g : 1 3 6
$G_{8,6,5,1}$	M : 0 1 2 3 6 7 8
	g : 0 1 3 6 7
$G_{9,4,3,1}$	M : 0 3 4 5 6 7 9
	g : 0 1 3 5 8
$G_{10,5,2,1}$	M : 0 2 3 5 7 9 10
	g : 1 4 6 7 8 9
$G_{11,8,3,2}$	M : 0 2 3 5 8 10 11
	g : 1 4 5 7 8 9 10
$G_{12,9,3,2}$	M : 0 3 4 5 6 8 12
	g : 3 4 5 8 9 11
$G_{13,6,5,2}$	M : 0 2 4 7 10 11 13
	g : 0 1 2 3 7 8 9 12
$G_{14,12,2,1}$	M : 0 3 5 6 8 9 11 13 14
	g : 1 3 11 12 13
$G_{15,8,6,5}$	M : 0 1 8 9 11 13 15
	g : 0 5 8 10 11 14
$G_{16,12,10,3}$	M : 0 1 2 8 13 15 16
	g : 0 1 3 5 9 13 14 15
$G_{17,10,7,6}$	M : 0 3 6 8 11 12 17
	g : 3 4 6 7 8 9 10 12 13 14 16
$G_{18,10,8,5}$	M : 0 1 2 3 4 10 11 12 13 15 18
	g : 1 4 7 9 11 12 14 15 17
$G_{19,12,10,4}$	M : 0 1 2 6 7 10 12 13 14 15 16 18 19
	g : 2 3 8 10 11 14 15 16 17 18
$G_{20,11,6,1}$	M : 0 1 2 5 7 9 11 13 14 15 17 19 20
	g : 0 1 3 8 9 11 17 18 19
$G_{21,17,8,3}$	M : 0 2 4 5 6 7 8 10 16 20 21
	g : 1 8 10 13 14 18 19 20
$G_{22,1}$	M : 0 1 5 6 7 9 10 12 13 14 15 16 18 19 22
	g : 0 3 6 8 14 16 19 21
$G_{23,5}$	M : 0 1 4 5 7 8 9 11 13 14 16 17 18 19 20 21 23
	g : 1 3 6 7 8 9 11 18 22
$G_{25,3}$	M : 0 1 6 9 11 14 16 18 19 23 25
	g : 0 3 7 9 11 12 13 14 21 24
$G_{28,3}$	M : 0 1 3 4 5 8 9 10 11 12 15 20 21 22 23 24 26 27 28
	g : 0 1 2 3 4 5 6 7 9 10 11 12 13 14 15 16 17 18 19 21 22 24 26 27
$G_{31,13}$	M : 0 1 2 3 5 7 8 9 11 12 13 14 16 17 18 19 22 27 28 30 31
	g : 1 4 8 9 13 15 19 24 26 28 30

Table 4.20 Figures of merit in dimensions $k = 2$ to 6 for the resulting generators

	$\rho^{(2)}$	$\rho^{(3)}$	$\rho^{(4)}$	$\rho^{(5)}$	$\rho^{(6)}$
$G_{5,3,2,1}$	6	4	4	4	3
$G_{6,5,2,1}$	7	5	3	3	3
$G_{7,5,3,1}$	8	6	6	5	4
$G_{8,6,5,1}$	9	6	5	5	5
$G_{9,4,3,1}$	10	8	6	6	6
$G_{10,5,2,1}$	11	8	8	7	6
$G_{11,8,3,2}$	12	9	8	8	8
$G_{12,9,3,2}$	13	9	9	8	7
$G_{13,6,5,2}$	14	10	10	9	9
$G_{14,12,2,1}$	15	11	10	9	8
$G_{15,8,6,5}$	16	12	11	11	9
$G_{16,12,10,3}$	17	12	12	11	11
$G_{17,10,7,6}$	18	14	13	12	12
$G_{18,10,8,5}$	19	15	13	13	11
$G_{19,12,10,4}$	20	16	14	14	12
$G_{20,11,6,1}$	21	16	15	14	12
$G_{21,17,8,3}$	22	18	16	15	12
$G_{22,1}$	23	17	17	15	13
$G_{23,5}$	24	16	15	15	15
$G_{25,3}$	26	20	19	17	15
$G_{28,3}$	29	24	23	18	18
$G_{31,13}$	32	24	24	22	19

4.2 LCGS USING MULTIPLICATIVE INVERSION

The inversive congruential generator is defined by

$$X_n = aX_{n-1}^{-1} + b \pmod{M},$$

where X^{-1} means the multiplicative inverse of an integer X modulo M and is defined as 0 if $X = 0$. In practice, M is a prime or a power of two, around 2^{30}. A random number in $[0, 1)$ is as usual defined by

$$u_n = \frac{X_n}{M}.$$

Obviously, the period of $u_n, n = 1, 2, ...,$ is at most M if M is a prime, and at most $M/2$ if M is a power of two.

There are two ways of computing the inverse of an invertible element X modulo M: the Euclidean algorithm and the square-and-multiply calculation using the Euler theorem $X^{-1} = X^{\phi(M)-1}$. For the latter approach, if $\phi(M)$ is of the form 2^l, then we have the following efficient algorithm:

Algorithm I for computing the multiplicative inverse of X modulo M

Step 1: $R = 1$
Step 2: for $i = 1$ to l $\{R = R^2 * X \pmod{M}\}$
Step 3: Output R as X^{-1}.

4.2.1 Generators Modulo 2^w

For the case $M = 2^w$, we have the following result [72]:

Theorem 4.7 *Let $M \geq 8$ be a power of two. Then the period of inversive congruential sequences with modulus M has a maximum period $M/2$ if and only if* $a = 1 \pmod{4}$, $b = 2 \pmod{4}$, *and* $X_0 = 1 \pmod{2}$.

Since in this case $\phi(M) = 2^{w-1}$, we have an efficient implementation using Algorithm I; that is, we have $X^{-1} = X^{2^{w-1}-1} \pmod{M}$. Figure 4.6 shows the

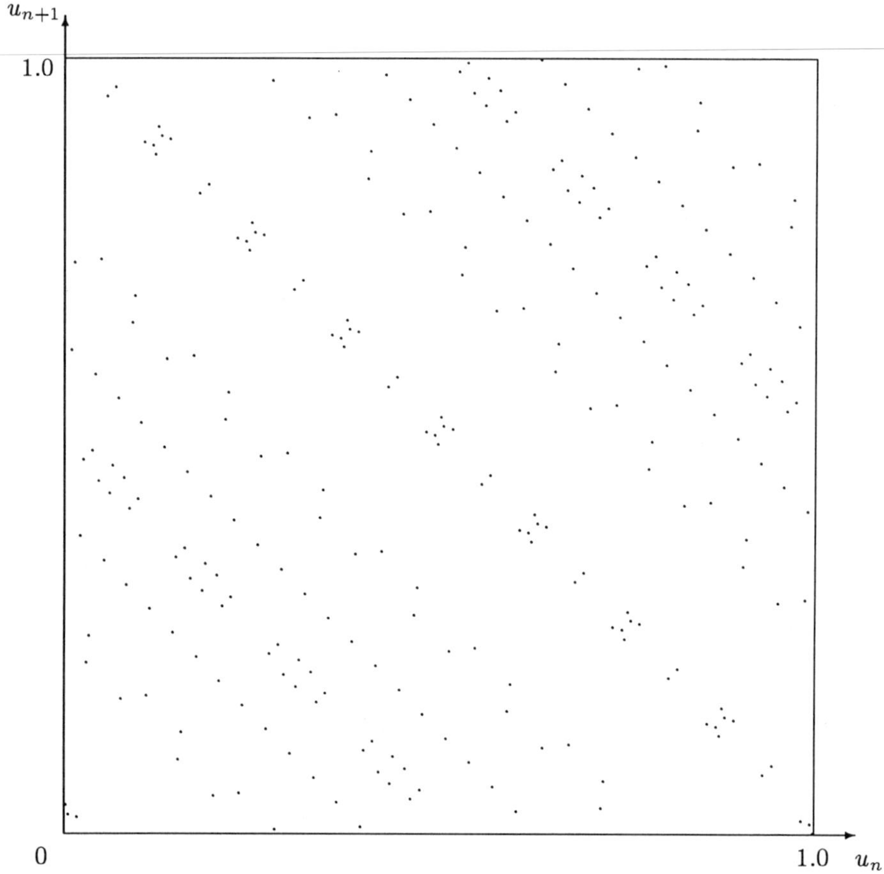

Figure 4.6 A set of two-dimensional points, $(u_n, u_{n+1}), n = 1, ..., 256$, produced by $u_n = X_n/512$ with $X_n = 9X_{n-1}^{-1} + 10 \pmod{512}$.

two-dimensional plot of $(u_n, u_{n+1}), n = 1, ..., 256$, over the entire period from the inversive congruential generator

$$X_n = 9X_{n-1}^{-1} + 10 \pmod{512}.$$

As the figure shows, there is a certain lattice structure in the plot. The lattice structure of inversive congruential sequences can be analyzed by generalizing the notion of potency, which was originally used for linear congruential sequences. In the same way as in Section 3.1, we consider the s-fold iterate of the difference operator $\Delta X_n = X_{n+1} - X_n$. We have

$$\Delta^s X_n = \sum_{j=0}^{s} (-1)^{s-j} \binom{s}{j} X_{n+j}.$$

As a measure of the lattice structure of sequences, we define 'potency' as the smallest integer s such that $\Delta^s X_n$ is constant modulo M. It is easy to see that this definition is a generalization of the potency for linear congruential sequences, because it is equivalent to the smallest integer s satisfying $(a-1)^s = 0 \pmod{M}$ when applied to them. We make the following conjecture:

Conjecture 4.1 *For maximum-period inversive congruential sequences with modulus $M = 2^w$ $(w \geq 3)$, the potency is always less than w.*

In fact, we have confirmed this conjecture up to $w = 12$ by exhaustive computations.

We should note that, in some cases, the potency s becomes infinite. For example, consider the periodic sequence $X_n, n = 1, 2, ...,$ modulo 5: 2,4,0,1,2,4,0,1,... In this case, by repeatedly differencing, we get

the first difference	$\Delta X_n, n = 1, 2, ...$:	2,1,1,1,2,1,1,1,2,...
the second difference	$\Delta^2 X_n, n = 1, 2, ...$:	4,0,0,1,4,0,0,1,4,...
the third difference	$\Delta^3 X_n, n = 1, 2, ...$:	1,0,1,3,1,0,1,3,1,...
the fourth difference	$\Delta^4 X_n, n = 1, 2, ...$:	4,1,2,3,4,1,2,3,4,...
the fifth difference	$\Delta^5 X_n, n = 1, 2, ...$:	2,1,1,1,2,1,1,1,2,....

Here, the fifth difference is the same as the first one. Thus, even for a periodic sequence, the potency is not guaranteed to be bounded.

Currently, little is known about the discrepancy of inversive congruential generators with modulus $M = 2^w$, except for the upper bound of the two-dimensional discrepancy over the entire period, namely, $D^{(2)}_{M/2} = O(M^{-1/2}(\log M)^2)$ [72].

4.2.2 Generators Modulo a Prime

Next, we consider the case in which the modulus M is a prime. We have a sufficient condition for the period to be maximum M.

Theorem 4.8 *If $x^2 - bx - a$ is a primitive polynomial over $GF(p)$, the period of inversive congruential sequences with a prime modulus $p \geq 5$ is equal to p.*

Since p is prime, we have $X^{p-1} = 1 \pmod{p}$. Thus, if p is a prime number of the form $2^l + 1$, then the multiplicative inverse of X modulo p is equal to $X^{2^l - 1}$, which means that we can use Algorithm I, given above for efficient implementation.

There is a result concerning the lattice structure of inversive congruential generators with prime moduli [27]:

Theorem 4.9 *Let $X_n, n = 0, 1, ...,$ be any maximum-period inversive congruential sequence with modulo p. Then for any $k \geq 2$, each hyperplane in $GF(p)^k$ contains at most k of the points $(X_n, X_{n+1}, ..., X_{n+k-1})$ with $\prod_{j=0}^{k-2} X_{n+j} \neq 0$, for $n = 0, 1, ..., p-1$.*

In Figure 4.7, we show the two-dimensional plot (u_n, u_{n+1}) over the entire period from the following generator:

$$X_n = 22X_{n-1}^{-1} + 10 \pmod{509}.$$

The figure implies that inversive congruential sequences seem to have more random-like distribution for prime moduli than for moduli of a power of two.

If p is a prime, we can use Legendre polynomials $h(x)$ so that $h(n) = X_n, n = 0, ..., p-1$, for any maximum-period inversive congruential sequence, namely,

$$h(x) = \sum_{n=0}^{p-1} (1 - (x-n)^{p-1}) X_n.$$

Thus, the potency s is at most the degree of $h(x)$, that is, at most $p-1$. In general, a polynomial $h(x)$ is called a permutation polynomial modulo p if $h(i) \neq h(j)$ for any pair (i, j) of integers with $0 \leq i \neq j \leq p-1$. We have the following result:

Proposition 4.13 *If an inversive congruential sequence with a prime modulus p has the maximum period p, then its potency is equal to the degree of the permutation polynomial associated with the sequence.*

Since it is known, see [72], that the degree of a permutation polynomial is at most $p-2$, we can combine this result with that obtained by Eichenauer [26]:

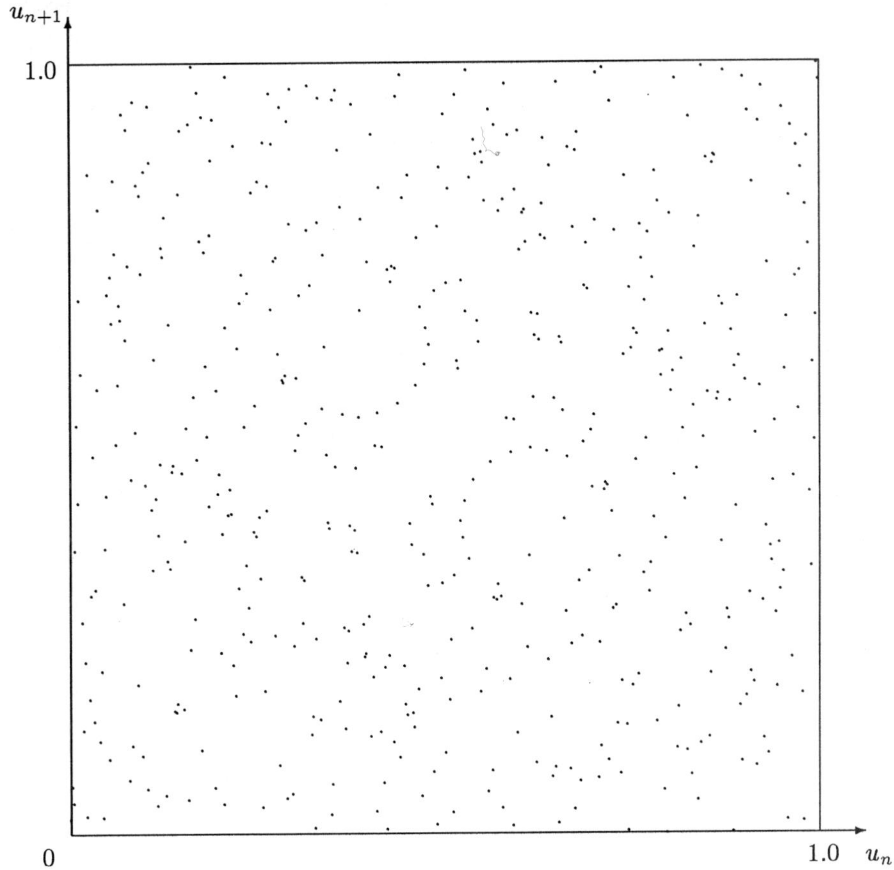

Figure 4.7 A set of two-dimensional points, $(u_n, u_{n+1}), n = 1, ..., 509$, produced by $u_n = X_n/509$ with $X_n = 22X_{n-1}^{-1} + 10 \pmod{509}$.

Proposition 4.14 *The potency s of maximum-period inversive congruential sequences with a prime modulus p satisfies*

$$\frac{p+1}{2} \leq s \leq p-2.$$

Note that a large potency does not necessarily mean that the sequence closely mimics (truly) random numbers. For example, consider the following sequence of period p:

$$0, 1, 2, ..., p-4, p-3, p-1, p-2, 0, 1, 2, ...$$

In this case, see [26], the potency is the maximal, $p-2$, but the sequence looks far from random.

Regarding the discrepancy of inversive congruential sequences, we have an upper bound due to Niederreiter [72]:

Theorem 4.10 *For maximum-period inversive congruential sequences with a prime modulus p, the k-dimensional discrepancy over the entire period is given as*

$$D_p^{(k)} = O(\frac{\log^k p}{p^{1/2}}).$$

Niederreiter [72] also obtained a lower bound which implies that the above theorem is the best possible up to a logarithmic factor.

Theorem 4.11 *Let (a, b) be the parameters for maximum-period inversive congruential sequences with a prime modulus $p \geq 5$. Then for a positive fraction, depending on p and t, of the set of parameters (a, b), we have*

$$D_p^{(k)} > \frac{t}{(2\pi + 4)p^{1/2}},$$

for all $k \geq 2$, where $0 < t < 1$.

Therefore, the two-dimensional discrepancy of inversive congruential sequences with prime moduli as well as with moduli of a power of two over the entire period roughly conforms with $O(p^{-1/2}(\log \log p)^{1/2})$, which is the order of the discrepancy for (truly) random sequences according to the law of the iterated logarithm. Anyway, the above discrepancy results are only for the entire period of the sequences. Since the entire period of the sequences is rarely used in practical applications, readers are strongly encouraged to explore the discrepancy bounds for only parts of the period.

4.3 RANDOM SEQUENCES IN CRYPTOGRAPHY

The unpredictability of sequences is an important topic in cryptography. Knuth [40] considered the predictability of linear congruential sequences. The problem was formulated as follows: Consider a linear congruential sequence $X_n, n = 1, 2, ...,$ from

$$X_n = aX_{n-1} + b \pmod{2^w}.$$

Let $X_n^{(h)}$ stand for the $h(\leq w)$ high-order bits of X_n. How can we determine the unknown constants a, b, and X_0 if w and $X_n^{(h)}, n = 1, 2, ..., N$, are given ? His result shows that the problem can be solved in $O(w^2)$ steps when $h \geq 2$; in other words, linear congruential sequences are found to be cryptographically insecure. This result was generalized to hold under several other different conditions, and finally the polynomial-time predictability of the sequences produced from general congruential generators was proved [43].

In this section, we discuss two types of "unpredictable" binary sequences. In both cases, however, we should notice that the results are only for the 'asymptotic' properties of finite sequences, i.e., if the sequences are very long.

4.3.1 Cryptographically Secure Sequences

First, we define the notion of a *pseudorandom bit generator* [13, 14, 110].

DEFINITION **4.12** A pseudorandom bit generator G is a sequence of polynomial-time algorithms $G_n, n = 1, 2, ...$ Each G_n stretches a (random) seed $s^{(n)}$ of length n into a pseudorandom sequence $x^{(l)}$ of length l, that is,

$$G_n : \{0,1\}^n \to \{0,1\}^l,$$

where $x^{(l)} = G_n(s^{(n)})$. We assume that G_n is evaluated in a polynomial time in n. Thus, the length l of $x^{(l)}$ is also polynomial in n. □

That is to say, in this section, we consider a set of pseudorandom bit generators, not one particular generator with one particular seed value, and consider the distribution of output sequences for each generator with seeds selected uniformly at random. Denote by $\nu_{R,l}$ the uniform distribution on the set of all l

bit sequences. Thus, we have $\nu_{R,l}(x^{(l)}) = 2^{-l}$. Let $\nu_{G,l}$ denote the probability distribution of l bit sequences generated from G_n for randomly selected $s^{(n)}$; we then have

$$\nu_{G,l}(x^{(l)}) = \frac{\#\{s^{(n)} : G_n(s^{(n)}) = x^{(l)}\}}{2^n}.$$

A pseudorandom sequence is usually defined by using two important concepts: unpredictability and indistinguishability. The notion of unpredictability, which is particularly important in cryptography, can be formalized through the following notion of a *next bit test* [13, 14, 110].

DEFINITION **4.13** A next bit test C is a sequence of probabilistic polynomial-time algorithms $C_l, l = 1, 2, \ldots$. Each C_l accepts pairs $(i, x^{(i)}), i = 1, 2, \ldots, l-1$, as input and produces one bit of output. □

By using this notion, we can now formalize unpredictability [13, 14, 110].

DEFINITION **4.14** A pseudorandom bit generator G is said to pass a next bit test C if for all polynomials $Q(l)$ and all sufficiently large l,

$$Pr[C_l(x^{(i)}) = x_{i+1}] < \frac{1}{2} + \frac{1}{Q(l)},$$

where x_i is the ith bit of $x^{(l)}$ drawn from $\nu_{G,l}$ for $i < l$. □

If a generator passes all next bit tests, it is said to be unpredictable to the right.

As stated above, another important notion concerning pseudorandom sequences is indistinguishability. Loosely speaking, this comes from the requirement that pseudorandom sequences can simulate, and should not be distinguished from, (truly) random sequences. The concept is formalized through the notion of a polynomial-time statistical test, defined as follows [13, 14, 110]:

DEFINITION **4.15** A statistical test T is a sequence of probabilistic polynomial-time algorithms $T_l, l = 1, 2, \ldots$, each of which accepts as input sequences $x^{(l)}, l = 1, 2, \ldots$, and produces one bit output. □

Indistinguishability is then defined as follows:

Definition **4.16** A pseudorandom bit generator G is said to pass a polynomial-time statistical test T if for all polynomials $Q(l)$ and all sufficiently large l,

$$|\mu_{G,l}^{T}(1) - \mu_{R,l}^{T}(1)| < \frac{1}{Q(l)},$$

where $\mu_{G,l}^{T}(1)$ and $\mu_{R,l}^{T}(1)$ denote the probability distributions for a test T to output 1 on input sequences drawn from distributions $\mu_{G,l}$ and $\mu_{R,l}$, respectively. □

If a generator passes all polynomial-time statistical tests, it is said to be indistinguishable (or cryptographically secure) from (truly) random sequences. Yao's theorem [110] links unpredictability and indistinguishability, and it has the following form:

Theorem 4.12 *A pseudorandom bit generator G passes all next-bit tests if and only if it passes all polynomial-time statistical tests.*

The immediate consequence of this theorem is that sequences produced from a generator which is unpredictable to the right are indistinguishable from (truly) random sequences. This implies that the generator is also unpredictable to the left.

Recently, it was proved that cryptographically secure generators exist if and only if one-way functions exist [37]. The definition of one-way functions is as follows:

Definition **4.17** Let $f_n : \{0,1\}^n \rightarrow \{0,1\}^{l(n)}, n = 1, 2, ...,$ be a sequence of functions, where $l(n)$ is monotone-increasing and computable in a polynomial time in n. Define $f(x) = f_n(x)$ if $x \in \{0,1\}^n$. A function f is said to be one-way if it satisfies the following conditions:

- f_n is computable in a polynomial time in n.
- For all sufficiently large n, all polynomials $Q(n)$, and all probabilistic polynomial-time algorithms A, the probability that, for an input $f_n(x)$, the algorithm A outputs an x' such that $f_n(x) = f_n(x')$ is smaller than $1/Q(n)$, where x is chosen according to $\mu_{R,n}$.

□

Unfortunately, however, it is still unknown whether one-way functions exist. We should notice that if $NP = P$ they do not exist, but $NP \neq P$ does not necessarily mean that they do exist.

There are several methods of constructing cryptographically secure sequences by using specific permutations believed to be one-way, which are based on discrete logarithm modulo a prime [14], quadratic residuasity modulo a composite integer (or integer factoring) [13], and RSA functions [40, 79]. We can summarize their specifications as follows:

- [discrete-log] Calculate $x_{i+1} = g^{x_i} \pmod{p}$ for $i = 1, 2, ...$, and if $x_{i-1} \leq (p-1)/2$ then $b_i = 1$; otherwise, $b_i = 0$, where p is a prime and g is a generator in $GF(p)$. Output $b_i, i = 1, 2, ...$
- [quadratic-residue] Calculate $x_{i+1} = x_i^2 \pmod{pq}$ for $i = 1, 2, ...$, and let b_i be the least significant bit of x_i, where p and q are distinct primes both congruent to 3 modulo 4. Output $b_i, i = 1, 2, ...$
- [RSA-function] Calculate $x_{i+1} = x_i^e \pmod{pq}$ for $i = 1, 2, ...$, and let b_i be the least significant bit of x_i, where p and q are distinct primes and $e > 2$ satisfies $\gcd(e, (p-1)(q-1)) = 1$. Output $b_i, i = 1, 2, ...$

4.3.2 Linear Complexity Profile

Consideration of whether a given individual finite sequence is random always leads to a contradiction, since the sequence resulting from n fair flips of a coin could yield any sequence of length n with equal probability. This suggests that, instead of individual sequences, we should consider whether a given subset of finite sequences can be regarded as random or not. In other words, we should find a nice characterization of sequences *randomly chosen* from all sequences of length n. Kolmogorov [41], and independently Chaitin [18], established a mathematical notion of the complexity of finite sequences. Before giving its definition, we need to briefly overview the definitions of the Turing machine and its relevant notions. Let S denote the state space with an initial state α and the set of final states Ω, and let the tape alphabet be denoted by $\mathcal{A} = \{0, 1, \#\}$, where $\#$ indicates a blank.

DEFINITION **4.18** A Turing machine is defined as a finite state machine with an infinite tape as a work space for the manipulation of symbols, which can be read and written via the tape head under the control of the machine by the commands $Z = \{ +1 \text{ (left)}, -1\text{(right)}, \text{and } 0\text{(stop)}\}$, where the next state function $\delta : S \times \mathcal{A} \times Z \to S \times \mathcal{A} \times Z$ also decides the next movement of the tape head and writes the output on the tape. Here we assume that the work tape is initially filled with blanks. □

We denote a Turing machine by $T = (S, \mathcal{A}, Z; \delta, \alpha, \Omega)$. The next result leads to the concept of a more general-purpose Turing machine [18].

Theorem 4.13 *There exists a Turing machine U such that for any Turing machine T there exists an input $P(T)$ to U, which is a binary program to make U simulate T.*

The Turing machine U is called the Universal Turing machine. Now, we are in a position to define the Kolmogorov complexity.

DEFINITION **4.19** T denotes a given Turing machine that generates a binary sequence $x^{(n)}$ of length n. Let $P(T)$ be the program with which the Universal Turing machine simulates T. Then we define the Kolmogorov complexity of a sequence $x^{(n)}$ by

$$\kappa(x^{(n)}) = \min\{|P(T)| \mid T \text{ generates } x^{(n)}\}.$$

□

The following result is essential [18, 41]:

Theorem 4.14 *For $i = 1, ..., n$, we have*

$$|\{x^{(n)} \in \{0,1\}^n \mid \kappa(x^{(n)}) \leq i\}| \leq 2^{i+1}.$$

This implies that most sequences of length n have Kolmogorov complexity close to n (the length of the sequence); that is, few sequences can have a description shorter than itself.

From the practical viewpoint, the Kolmogorov complexity is of little use, because we cannot easily calculate the minimum size of a program for the Universal Turing machine that generates a given sequence. The most popular concept in practical use in the field of cryptography for the complexity of finite sequences is called *linear complexity*, which is defined as follows [82]:

DEFINITION **4.20** Let $x_1, x_2, ..., x_n$ be a binary sequence of length n. The linear complexity $\lambda(x^{(n)})$ of the sequence is defined by the minimum degree of a polynomial, $f(z) = z^r + a_{r-1}z^{r-1} + \cdots + a_1 z + a_0$, over $GF(2)$ with $r < n$, such that

$$x_{i+r} = a_{r-1}x_{i+r-1} + \cdots + a_1 x_{i+1} + a_0 x_i \pmod{2},$$

for $i = 1, ..., n-r$. □

It is known [82] that the Berlekamp-Massey algorithm computes the linear complexity of any binary sequence of length n in $O(n^2 \log n)$ bit operations.

For example, consider a binary sequence $x_1, x_2, ...,$ of period T. For the first n elements of the sequence, the linear complexity is at most n if $n < T$; otherwise it is at most T. Rueppel [82] showed the following:

Theorem 4.15 *The average value of the linear complexity over all binary sequences of length n is $\frac{n}{2} + c_n$, where $0 \le c_n \le \frac{5}{18}$ and its variance is about $\frac{86}{81}$.*

The next theorem [11] gives an important relation between the Kolmogorov complexity and the linear complexity:

Theorem 4.16 *For all $0 < \epsilon < 1$, we have*

$$Pr[x^{(n)} \in \{0,1\}^n \mid (1-\epsilon)\lambda(x^{(n)}) \le \frac{\kappa(x^{(n)})}{2} \le (1+\epsilon)\lambda(x^{(n)})] \to 1$$

when $n \to \infty$.

The main consequence of this theorem is that the linear complexity can be viewed as a practical alternative to the Kolmogorov complexity for investigating the randomness of finite sequences.

Now, we introduce the notion of a linear complexity profile [82]. Denote by $x^{(n)}$ the first n elements of a sequence $x_1, x_2, ...$

DEFINITION **4.21** The sequence $\lambda(x^{(n)}), n = 1, 2, ...,$ is called the linear complexity profile of $x_1, x_2, ...$ □

DEFINITION **4.22** A (binary) sequence $x_1, x_2, \ldots,$ in $GF(2)$ is said to have a perfect linear complexity profile if it has a linear complexity profile $\lambda(x^{(n)}) = [(n+1)/2]$ for all $n = 1, 2, \ldots$. □

The above results raise the question of how to obtain a sequence that has a perfect linear complexity profile. This question was first answered by Wang and Massey (see [82]), although in fairness it should be said that Baum and Sweet [9] had already reached the same result under different circumstances.

Theorem 4.17 *A sequence $x_1, x_2, \ldots,$ in $GF(2)$ has a perfect linear complexity profile if and only if the sequence satisfies*

$$x_{2i+1} = x_{2i} + x_i \quad (\text{mod} 2), \ \textit{for } i = 1, 2, \ldots, \tag{4.27}$$

with $x_1 = 1$.

We should note that all the elements with even suffixes, namely, $x_{2i}, i = 1, 2, \ldots,$ can be arbitrarily chosen. Thus, for the first $2n$ elements of such sequences, there are 2^n possible choices. Anyway, the result means that perfect linear complexity profile sequences are not suitable for cryptographic purposes, because of their simple linear dependent relation, given in (4.27). In fact, it is unlikely that independent random samples from a certain probability distribution always take the mean value of that distribution. They should properly fluctuate above and below the mean. Niederreiter [71] investigated the distribution of linear complexity profiles of (truly) random binary sequences.

Finally, we give an interesting relation between a perfect linear complexity profile and Fibonacci polynomials [71]:

Theorem 4.18 *A sequence $x_1, x_2, \ldots,$ in $GF(2)$ has a perfect linear complexity profile if and only if the generating function $S(z) = \sum_{i=1}^{\infty} x_i z^{-i}$ is irrational and has a continued fraction expansion*

$$S(z) = [0; A_1(z), A_2(z), \ldots],$$

such that $\deg(A_i) = 1$ *for all $i \geq 1$.*

It is easy to see that each of the rational convergents $P_i(z)/Q_i(z), i = 1, 2, \ldots,$ for $S(z)$ corresponds to a pair of Fibonacci polynomials $(F_{i-1}(z), F_i(z))$.

* * *

Notes and Remarks

Section 4.1 dealt with important results obtained in the past decade on shift-register random number generators. In his book [72], Niederreiter devotes one chapter to the same topic more theoretically, in particular, from the viewpoint of discrepancy. Tezuka [93] discusses the $LS(q)$ sequence more generally. Though we defined only Fibonacci polynomials over $GF(2)$, the general definition of Fibonacci polynomials over rationals dates back to the 1960's [12].[2]

General results on nonlinear congruential generators in $GF(p)$, p prime, with maximal periods were obtained by Eichenauer and Lehn [25] and by Niederreiter [72].

Among the many good publications on the practical use and implementation of random numbers for cryptography, we recommend the Proceedings of the Crypto and Eurocrypt conferences (see Lecture Notes in Computer Science, published by Springer-Verlag).

[2] I am indebted to Professor Peter J.-S. Shiue for drawing my attention to the reference [12] and others relevant to Fibonacci polynomials.

5

STATISTICAL TESTS

Every random number generator will fail
in at least one application.
– Donald E. Knuth (1969)

As we have already seen, theoretical tests such as the spectral test give global (structural) information mainly about the distribution properties of the entire period of linear congruential sequences as well as $LS(2)$ sequences, whereas the discrepancy test is able to provide upper bounds for the deviation of a part of the period of these sequences from the uniform distribution. Although the concept of discrepancy is applicable in general to any pseudorandom sequences, it is difficult to obtain the exact value even for simple generators such as linear congruential sequences, except for the two-dimensional discrepancy of a small number of points. For other generators, it becomes more difficult to obtain even upper bounds for discrepancies in low dimensions.

As the period length becomes very large, we need to look at not only global information about the entire period of sequences but also the local distribution of parts of sequences, for which empirical tests are considered to be the last resort in practice. Empirical tests are of a purely statistical nature. They compare any given sequences with (truly) random sequences from the statistical viewpoint, and detect the non-uniformity and specific structures of dependency between the numbers in the sequences. The main advantage of empirical statistical tests is that they can be applied to any sequences from any generators for any part of sequences. In particular, they provide local information on pseudorandom sequences; that is, they are applicable to local segments of sequences. On the other hand, empirical tests cannot be applied to the entire period of sequences, because the period lengths of pseudorandom sequences in practical use are too large to make empirical tests feasible within a reasonable computing time.

In his book [40], Knuth proposed a battery of statistical tests, the so-called standard tests for random number generators. However, today these standard tests have become totally insufficient; that is, as the scale of computer simulation becomes ever larger with the evolution of computing power, it turns out that some generators that passed these tests often produce significant biases when they are used in large-scale simulations. In this context, Marsaglia [57] proposed a new set of stringent tests. However, as we showed before in Chapter 3, even AWC and SWB generators, which pass all these stringent tests, suffer from bad integer lattice structures and perform badly in dimensions higher than the degree of their recurrence relations. In fact, it has recently been found [28] that these generators produce significant biases in sophisticated applications such as high-precision Ising model simulation.

At the time of writing (and maybe at any time in the future), we cannot provide a "sufficient" set of empirical tests or "powerful" theoretical tests for uniform random numbers, although of course, we know that no pseudorandom sequences arithmetically produced can survive all empirical tests. So, from the engineering viewpoint, our best strategy in investigating pseudorandom numbers will be to combine the knowledge as much as possible from both theoretical tests and empirical tests, because of their complementary nature.

5.1 DESCRIPTION OF TEST PROCEDURES

In general, we set the null hypothesis H_0: "The sequence is a sample of uniformly and independently distributed random variables in $[0,1)$," and apply a statistical test to the sequence to find empirical evidence against H_0. We can use any function of a finite number of uniform random variables in $[0,1)$ as a statistic T, a test for H_0, provided that its theoretical distribution under H_0 is known. Usually, we apply a given test to N distinct segments of the sequences to obtain $T_1, ..., T_N$ as the values of test statistics.

5.1.1 Tests for Goodness of Fit

First, we describe tests for goodness of fit for comparing the empirical distribution of the values $T_1, ..., T_N$ with the theoretical distribution of T under H_0. The most popular are the chi-squared test and the Kolmogorov-Smirnov (K-S) test.

chi-squared test

The chi-squared test is suitable for the case in which the theoretical distribution is discrete. Assume that the statistic T takes on n possible values $v_1, ..., v_n (n > 1)$. From the N independent observations, we accumulate the number of outcomes f_i that take on v_i for each i. Let $\bar{f}_i$ be the expected number of outcomes out of N total ones for each i. If the hypothesis H_0 is true, then it is known that the statistic

$$\chi^2_{n-1} = \sum_{i=1}^{n} \frac{(f_i - \bar{f}_i)^2}{\bar{f}_i}$$

has asymptotically the chi-squared distribution with $n-1$ degrees of freedom. The approximation becomes reasonable if $\min_{1 \le i \le n} \bar{f}_i \ge 5$. The greater n becomes, the greater the power of the test. However, when n is too large, we need such a large value of N, the number of observations, that the test becomes infeasible because of practical limitations of computing power and memory.

The probability density function of the chi-squared distribution with n degrees of freedom is given by

$$f(x) = \frac{e^{-x/2} x^{n/2-1}}{2^{n/2}\Gamma(n/2)} \quad \text{for } x \ge 0.$$

It is known that for $n > 30$, the formula $Y = \sqrt{2\chi^2_n} - \sqrt{2n-1}$ closely approximates $N(0,1)$, which stands for the normal distribution with mean 0 and variance 1. As a more accurate approximation formula, we have

$$Y = \sqrt{\frac{9n}{2}} \left(\left(\frac{\chi^2_n}{n} \right)^{1/3} - \left(1 - \frac{2}{9n}\right) \right)$$

for $n > 10$. Roughly speaking, if the value of χ^2_n is large, we judge the sequence to be non-random (the hypothesis H_0 is rejected.)

Kolmogorov-Smirnov test

This test works well for continuous distributions. The steps of the test are as follows: Let $F(x)$ be the theoretical cumulative distribution function for the test T. First, we sort the N observations, $T_1, ..., T_N$, so that $T_i \leq T_{i+1}$ for $i = 1, ..., N-1$, and then compute

$$K_N^+ = \sqrt{N} \max_j \left(\frac{j}{N} - F(T_j) \right), \quad j = 1, 2, ..., N,$$

$$K_N^- = \sqrt{N} \max_j \left(F(T_j) - \frac{j-1}{N} \right), \quad j = 1, 2, ..., N,$$

which are the maximum positive and negative differences between the empirical cumulative distribution and the theoretical one. We can compare the values of K_N^+ and K_N^- with the theoretical values tabulated for various values of N found in Knuth [40] or many standard statistics texts. Unlike in the chi-squared test, which is an approximation and is applicable only when the number of observations is large, those tabulated values for the K-S test are exact for any value of N. Furthermore, when $N \geq 1000$, we have a good approximate distribution of K_N^+ or K_N^-, given by $F(x) = 1 - e^{-2x^2}$. Therefore, we can compute the threshold value z such that $Pr[K_N^\pm > z] = \alpha$, $0 \leq \alpha \leq 1$, namely $z \approx \sqrt{\frac{\log \alpha}{-2}}$.

5.1.2 Specific Tests

As Marsaglia [57] said, "Anyone with a knowledge of probability theory should be able to create his own tests", in principle, there is no limit to the number of empirical tests that can be designed for random number generators. In the following subsections, we describe some typical specific tests in practical use, giving them in alphabetical order. Here, we denote by $\left\{ {n \atop k} \right\}$ Stirling numbers of the second kind, namely, the numbers of ways of partitioning n (different) objects into k parts. We have $\left\{ {n \atop k} \right\} = k\left\{ {n-1 \atop k} \right\} + \left\{ {n-1 \atop k-1} \right\}$ with $\left\{ {n \atop n} \right\} = \left\{ {n \atop 1} \right\} = 1$. In the following, we denote by $U_1, U_2, ...,$ a sequence of observations of uniform random variables in $[0, 1)$.

autocorrelation test

Given n observations $U_1, ..., U_n$, we define the autocorrelation with delay d, denoted by $AC(d)$, as

$$AC(d) = \frac{1}{n-d} \sum_{i=1}^{n-d} \left(U_i - \frac{1}{2} \right) \left(U_{i+d} - \frac{1}{2} \right).$$

Under H_0, we have $E(AC(d)) = 0$ and $Var(AC(d)) = \frac{1}{144(n-d)}$. Thus, each set of n observations produces each value of the test statistics $T = 12\sqrt{n-d}AC(d)$, whose distribution becomes asymptotically the normal distribution $N(0,1)$ when n is much larger than d.

birthday spacings test

Given n observations $U_1, ..., U_n$, let $d > 1$ be an integer. Sort a sequence of integers $[dU_1], ..., [dU_n]$ in $[0, d-1]$ to get $I_1 \leq I_2 \leq \cdots \leq I_n$. Let Y be n minus the number of distinct spacings among $I_2 - I_1, I_3 - I_2, ..., I_{n-2} - I_{n-1}$, and $I_1 + d - I_n$. If d is large enough, Y is known to have an asymptotically Poisson distribution, whose probability density function is

$$f(i) = \frac{\mu^i e^{-\mu}}{i!} \quad \text{for } i = 0, 1, ...$$

with mean $\mu = n^3/4d$. Hence, by independently repeating this procedure N times, we can apply the chi-squared test to the resulting N test statistics.

collision test

By grouping nk observations $U_1, ..., U_{nk}$, we obtain n non-overlapping k-tuples $(U_{ki+1}, ..., U_{k(i+1)})$. Consider the equal division of the k-dimensional unit hypercube into d^k disjoint subcubes, each of which has a volume of $1/d^k$. In this setting, as the dimension becomes high, the number of subcubes becomes huge. This test is suitable for such situation. We say that a collision occurs if a k-tuple falls into an occupied cell. The number of collisions is given as follows:

$$C = n - \#(\text{occupied cells}).$$

We have

$$Pr[C = c] = \frac{m \cdots (m - n + c + 1)}{m^n} \left\{ \begin{array}{c} n \\ n - c \end{array} \right\},$$

where $m = d^k$. By independently repeating this procedure N times, we can apply the chi-squared test to the resulting N test statistics.

coupon collector's test

Produce Y observations $U_1, ..., U_Y$, where Y is the smallest integer such that a sequence of integers $[dU_1], ..., [dU_Y]$ contains all d integers in $[0, d-1]$. We have

$$Pr[Y = k] = \frac{d!}{d^k} \left\{ \begin{array}{c} k-1 \\ d-1 \end{array} \right\},$$

for $k \geq d$. By independently repeating this procedure N times, we can apply the chi-squared test to the resulting N test statistics.

frequency test

Given N observations $U_1, ..., U_N$, we divide the interval $[0, 1)$ into n disjoint subintervals $[0, a_1)$,$[a_1, a_2)$,..., and $[a_{n-1}, a_n)$ with $a_n = 1$. Under the hypothesis H_0, we have the following theoretical distribution of the expected number $\bar{f}_i$ of observations falling into the i-th subinterval $[a_{i-1}, a_i)$, out of the total of N observations:

$$\bar{f}_i = N(a_i - a_{i-1}).$$

By letting f_i be the actual number of observations in the ith interval, we can apply the chi-squared test directly to these data.

gap test

Given N observations $U_1, ..., U_N$, we regard the sequence as a periodic cycle of N. Specify a range $[a, b)$, where $0 \leq a < b < 1$, and define the gap length as the length of consecutive observations falling in the range. Under H_0, we have the geometrical distribution

$$p_i = Pr[\text{the gap length is } i] = (b - a)^i(1 - b + a),$$

and

$$p'_i = Pr[\text{the gap length is } i \text{ or larger}] = (b - a)^i.$$

Suppose that there are n gaps in a cycle. For $i = 0, ..., k - 1$, the expected number of gap length i is np_i, and the expected number of gaps whose length is greater than or equal to $k - 1$ is np'_{k-1}. We can apply the chi-squared test to them.

maximum of t test

Given k observations $U_1, ..., U_k$, find the maximum of the k observations. Under H_0, the maximum has the distribution

$$Pr[\text{the maximum} < x] = x^k.$$

By independently repeating this procedure N times, we can apply the K-S test to the resulting N test statistics.

Table 5.1 Table for μ and σ for three sets of d and n

d	n	μ	σ
10	2^{21}	141,909	290.26
11	2^{22}	1,542,998	638.75
11	2^{23}	567,637	580.80

nearest pair test

Given nk observations $U_1, ..., U_{nk}$, treat nk observations as n k-dimensional points $(U_{ik+1}, ..., U_{ik+k}), i = 0, 1, ..., n-1$. Let D be the Euclidean distance between the nearest pair of these n points. Under H_0, if n is large enough, the distribution of the test statistic $x = D^k$ becomes approximately the exponential distribution, whose probability density function is

$$f(x) = \frac{e^{-x/\mu}}{\mu} \quad \text{for } x \geq 0$$

with mean $\mu = 2/\kappa_k n^2$. By independently repeating this procedure N times, we can apply the K-S test to the resulting N test statistics.

OPSO test

The overlapping-pairs-sparse-occupancy (OPSO) test proposed by Marsaglia [57] looks like the collision test except for that the former deals only with two dimensions and counts empty cells instead of collisions. Given n observations $U_1, ..., U_n$, consider the pairs $([2^d U_i], [2^d U_{i+1}]), i = 1, 2, ..., n-1$, and $([2^d U_n], [2^d U_1])$. If we regard each pair as an array index for $(l, m), 0 \leq l, m < 2^d$, then it is known that the number of empty array indices has a normal distribution, with the values of μ and σ given in Table 5.1 due to Marsaglia, if n is large enough. By independently repeating this procedure N times, we can apply the K-S test to the resulting N test statistics.

permutation test

Given k observations $U_1, ..., U_k$, we obtain a permutation of the indices $\{1, ..., k\}$ by sorting the observations. In accordance with the independence of random numbers, we have

$$Pr[\text{the permutation is the same as the prespecified one}] = \frac{1}{k!}.$$

By independently repeating this procedure N times, we can apply the chi-squared test to the resulting N test statistics.

poker test

This test is an analogue of the card game Poker. As Birthday Spacings Test, we consider a sequence of integers $[dU_1], ..., [dU_{Nk}]$ in $[0, d-1]$. Divide this sequence into N sets of k consecutive integers. The event $E(i), 1 \leq i \leq k$, means that each set contains exactly i distinct integers. Then we have

$$Pr[E(i) \text{ occurs}] = \frac{d(d-1)\cdots(d-i+1)}{d^k} \left\{ \begin{array}{c} k \\ i \end{array} \right\},$$

to which the chi-squared test is applied.

run test

Given observations $U_1, U_2, ...$, we define up-runs as a set of monotone increasing segments. For example, consider the following sequence:

$$0.012,\ 0.345,\ 0.678,\ 0.9, \mid 0.876, \mid 0.543, \mid 0.21, \mid 0.123,\ 0.456,$$

where the vertical lines show the ends of up-runs of the sequence. This particular sequence has one up-run of length 4, one up-run of length 2, and three up-runs of length 1. Down-runs are defined similarly. As easily seen, there is a dependency between successive run lengths. A practical approach, due to Knuth, to remove the dependency is discarding the number immediately following a run. For the above sequence, we have

$$0.012,\ 0.345,\ 0.678,\ 0.9, \mid 0.543, \mid 0.123,\ 0.456,$$

which now has one up-run of length 4, one up-run of length 2, and one up-run of length 1, where all these run lengths are independent. For such a modified sequence, in general, it is known that

$$Pr[\text{the length of up-runs (or down-runs) is } k] = \frac{k}{(k+1)!}$$

for $k = 1, ..., m-1$, and

$$Pr[\text{the length of up-runs (or down-runs) is } m \text{ or larger}] = \frac{1}{m!}.$$

Thus, by using this probability distribution, we can apply the usual chi-squared test with $m-1$ degrees of freedom.

serial test

By grouping the Nk observations $U_1, ..., U_{Nk}$, we obtain N non-overlapping k-tuples $(U_{ki+1}, ..., U_{k(i+1)})$. Consider the equal division of the k-dimensional unit hypercube into d^k disjoint subcubes, each of volume $1/d^k$. Under the hypothesis H_0, the expected number $\bar{f}_i$ of k-tuples falling into each subcube is

$$\bar{f}_i = \frac{N}{d^k}$$

for the total of N k-tuples. As in the uniformity test, we can test these statistical values on the basis of the chi-squared test.

5.1.3 Multi-Level Tests

We call a test to obtain one value of the test statistic T a one-level test. The specific tests described above are all one-level tests. The chi-squared test and the Kolmogorov-Smirnov (K-S) test based on N values of one-level statistics are called two-level tests. We can perform a three-level test by repeating two-level tests R times for distinct sets of $(T_1^{(i)}, ..., T_N^{(i)})$ for $i = 1, ..., R$, from different parts of the sequence to obtain the values of the statistics $D_1, ..., D_R$. Then we may apply the K-S test again to compare the empirical distribution of these K-S or chi-squared statistics with their theoretical distributions. A simpler approach is the use of the descriptive level δ, defined as

$$\delta(D_i) = Pr[z < D_i \mid H_0].$$

Since $\delta(D_i)$ should be uniformly distributed in $[0, 1)$ under H_0, we can compare the distribution of the R values of $\delta(D_i)$ with the uniform distribution in $[0, 1)$ by using the K-S test.

Tests with four levels or more can be defined in a similar way. As the number of levels becomes higher, the test will be more powerful. However, the main problem of higher-level tests is that their exact theoretical distributions (even for the first level) under H_0 are not available in most cases, for which we usually use the asymptotic approximate distribution. Thus, the higher the level, the greater the degree of approximation; that is, the less reliable the test results. Two- or three-level test will be the best compromise.

5.2 TESTS USING NON-UNIFORM RANDOM VARIATE GENERATION

Generally, non-uniform random number generation can be said to be the transformation of a finite set of uniform variables into a set of variables having a target distribution. Thus, we can use any of such transformations as a statistical test under the H_0 hypothesis. Among many types of probability distributions in practical use, the variates from normal distributions are most frequently used, because of their central importance in the fields of statistics and probability theory, and the Box-Muller method is one of the most common transformation techniques for generating such variates. Because of its simplicity, this method has recently been more and more frequently used in computer simulations.

In 1973, Neave [67] pointed out that the combination of the Box-Muller method and the conventional linear congruential method produces undesirable results in some cases. To be more precise, the tail distribution generated by this combination differs markedly from the true distribution. This result is significant because in certain applications such as rare event simulations, the tail distribution should be exact. One consequence is that we have to be very careful in choosing the parameters of linear congruential sequences for the Box-Muller method. In this section, which is based on [90], we will describe the details of the Neave effect, which possibly occurs not only with linear congruential sequences but also with $LS(2)$ sequences.

5.2.1 Box-Muller Method and Neave Effect

First, recall that the Box-Muller method is given by

$$\begin{aligned} V_1 &= \sqrt{-2\log U_1}\sin(2\pi U_2), \\ V_2 &= \sqrt{-2\log U_1}\cos(2\pi U_2), \end{aligned}$$

where U_1 and U_2 are independent uniform random variates in $[0,1)$. Then V_1 and V_2 are independent normal random variates from $N(0,1)$. Note that

U_1 should not take the value of zero to avoid the overflow of $\log 0 = -\infty$. For a 32-bit word-size computer, $|V_1|$ and $|V_2|$ are bounded from above by $\sqrt{-2\log 2^{-32}} \approx 6.660$.

Neave [67] pointed out that some linear congruential sequences interact adversely with this Box-Muller method. Let the linear congruential sequence be

$$X_n = aX_{n-1} \pmod{M}.$$

The Box-Muller method is then given as

$$\begin{aligned} V_1 &= \sqrt{-2\log\frac{X_n}{M}}\sin(2\pi\frac{aX_n}{M}), \\ V_2 &= \sqrt{-2\log\frac{X_n}{M}}\cos(2\pi\frac{aX_n}{M}). \end{aligned}$$

As shown in [16, p. 223], the plots of (V_1, V_2) lie on a spiral. Hereafter, we deal only with V_1 unless otherwise specified, because similar results hold for V_2. Neave obtained the approximate range of V_1 with respect to the size of multipliers for linear congruential sequences; that is, the range is given as

$$\left[-\sqrt{2\log\frac{4a}{3}}, \sqrt{2\log 4a}\right] \tag{5.1}$$

for V_1. This implies that the smaller the value of a, the narrower the corresponding range. Specifically, when a is around $\sqrt{M} \approx 2^{16}$, V_1 takes values only in the approximate range -4.5 through 4.5. Note that multipliers with a size of around $\sqrt{M}$ are very often employed, due to their portability in implementation, to avoid overflow at multiplications with single-precision modular arithmetic. Since the probability of the true normal random variates falling within the tails, $[-\infty, -4.5]$ and $[4.5, \infty]$, is around 6×10^{-6}, we can say that simulation using more than 10^6 normal variates should not use the Box-Muller method with such generators.

In Figure 5.1, we show the tail distribution for the entire period $2^{31} - 2$ of the sequence generated by the most popular generator GGL. The range was exhaustively calculated as $[-4.476239, 4.717016]$, which turned out to be almost identical with the approximate range $[-4.475, 4.715]$ obtained from (5.1). Table 5.2 gives the detail of the distribution in the ranges $[-3.7, -3.6]$ and $[3.7, 3.8]$, where we should note that the 'Deviation' asymptotically follows $N(0, 1)$. An important observation here is that a significant disparity starts from around $|V_1| = 3.6$, far ahead of the bound at which the tail disappears. Neave explained

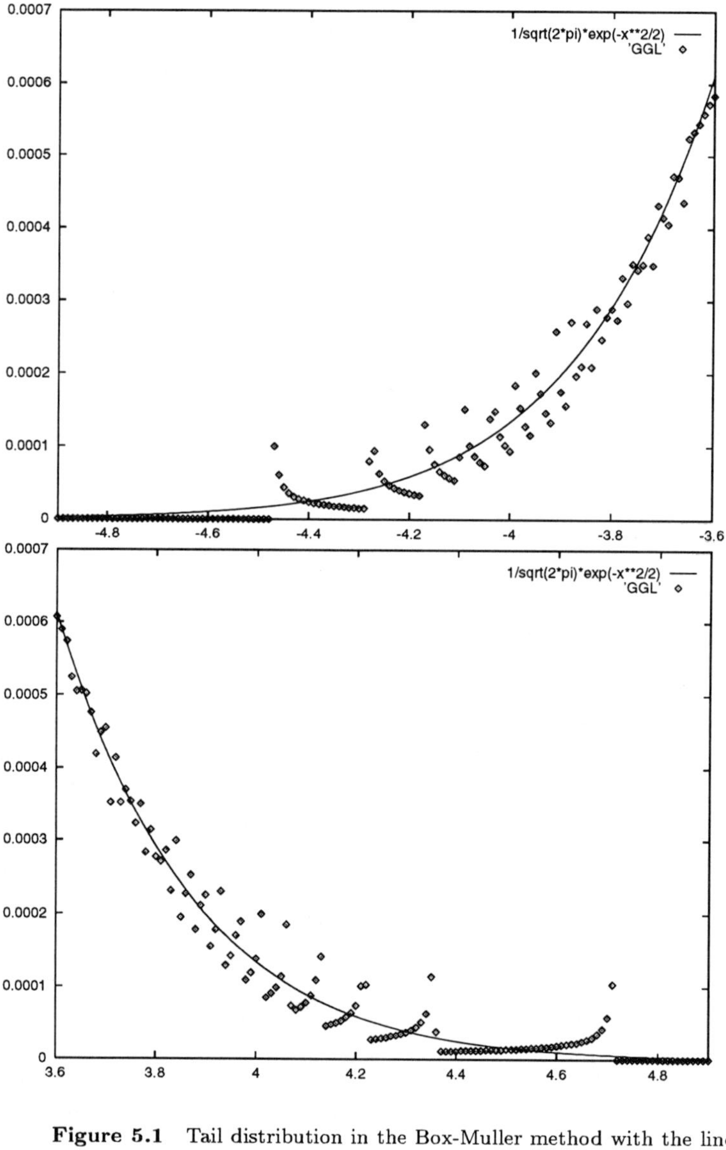

Figure 5.1 Tail distribution in the Box-Muller method with the linear congruential sequence, $X_n = 16807X_{n-1} \pmod{2^{31} - 1}$, over the entire period.

Table 5.2 Frequencies of V_1 in the tail generated by the Box-Muller method with the linear congruential sequence, $X_n = 16807X_{n-1} \pmod{2^{31} - 1}$, over the entire period

Range	Observed O	Expected E	Deviation $(O-E)/\sqrt{E}$
−3.69 : −3.70	8729	9298	−5.9
−3.68 : −3.69	10126	9642	4.9
−3.67 : −3.68	10089	10007	0.8
−3.66 : −3.67	9357	10372	−9.9
−3.65 : −3.66	11244	10758	4.6
−3.64 : −3.65	11447	11166	2.6
−3.63 : −3.64	11692	11574	1.0
−3.62 : −3.63	11974	12004	−0.2
−3.61 : −3.62	12278	12455	−1.5
−3.60 : −3.61	12552	12906	−3.1
3.70 : 3.71	9772	8955	8.6
3.71 : 3.72	7559	8632	−11.5
3.72 : 3.73	8887	8312	6.3
3.73 : 3.74	7566	8010	−4.9
3.74 : 3.75	7937	7715	2.5
3.75 : 3.76	7608	7430	2.0
3.76 : 3.77	6929	7157	−2.6
3.77 : 3.78	7507	6893	7.3
3.78 : 3.79	6080	6635	−6.8
3.79 : 3.80	6752	6390	4.5

the disparity as due to the discontinuities that occur at the zeros of the equation

$$\frac{dZ}{dX} = 0,$$

where $Z = \sqrt{-2\log X}\sin(2\pi a X)$.

Note that the Neave effect occurs even when we use the sequence in reverse order, that is,

$$\begin{aligned} V &= \sqrt{-2\log\frac{X_n}{M}}\sin(2\pi\frac{X_{n-1}}{M}) \\ &= \sqrt{-2\log\frac{X_n}{M}}\sin(2\pi\frac{a^{-1}X_n}{M}), \end{aligned}$$

where $a^{-1}a = 1 \pmod{M}$, provided that the size of a^{-1} is small (even if the size of a is large enough). The next example is from the classroom generator proposed by Marsaglia and Zaman, which was discussed in Example 3.1. We consider the sequence

$$X_n = (6^{-1})^{21}X_{n-1} \pmod{M},$$

where $M = 6^{21} + 6^2 - 1$. Let $u_n = X_n/M$, and suppose that the Box-Muller method is implemented in the following way:

$$V_1 = \sqrt{-2\log u_{4n}}\sin(2\pi u_{4n-2}),$$

and

$$V_2 = \sqrt{-2\log u_{4n-1}}\cos(2\pi u_{4n-3}).$$

Since $6^{42} = (6^2-1)^2 = 1225 \pmod{M}$, we can also observe the Neave effect for this case.

5.2.2 Box-Muller Method with $LS(2)$-Sequences

As Ripley states [78, p.59], "the best way to understand the Neave effect is to note that we can be concerned with large values of V_1 and V_2, and thus with small values of U_i". This observation also holds for $LS(2)$ sequences; that is to say, when the degree of $g(z)$ is small, the values of V_1, which are generated in the same way as above by replacing linear congruential sequences with $LS(2)$ sequences, fall within the bounded range, as in the Neave effect. These assertions can be made more quantitatively. Denote $u_n(z) = f_n(z)/M(z)$, i.e., $u_n = \sigma(u_n(z))$. Define $B(t) = \sqrt{-2\log(2^{-d-1}t)}\sin(2\pi t)$ for $t > 0$, where $d = \deg(g(z))$. Since $2^d u_n \le \sigma(g(z)u_n(z)) < 2^{d+1}u_n$, we have the following bounds for $V_1 = \sqrt{-2\log u_n}\sin(2\pi u_{n+1})$:

$$\inf_{t>0} B(t) \le V_1 \le \sup_{t>0} B(t) \tag{5.2}$$

for any (u_n, u_{n+1}). Numerical computation [90] shows that this bound closely approximates the actual range of V_1 for $1 \le d \le 15$. Similar arguments to Neave's analysis on the discontinuities for linear congruential sequences can also be applied to the case of $LS(2)$ sequences.

Now, we give two practical examples in which a Neave-like effect occurs with $LS(2)$ sequences. The first example is a component of the combined generator, SUPER-DUPER, already described in Example 4.6 of Section 4.1. Figure 5.2 shows the tail distribution of V_1 with this sequence over the entire period $(2^{21}-1)(2^{11}-1)$. Clearly, a phenomenon similar to the Neave effect occurs. Exhaustive calculation shows that for this generator the range into which V_1 falls is $[-4.622979, 4.856391]$, while the bound for V_1 from (5.2) gives the range $[-4.770, 4.996]$ for $\deg(g(z)) = 15$, which can be seen as a good approximation. Table 5.3 gives details of the distribution in the range $[-3.95, -3.85]$ and $[3.65, 3.75]$, which show that the disparity has already started at around $|V_1| = 3.7$.

The second example includes two $LS(2)$ sequences, whose parameters are $M(z) = z^{127} + z^s + 1$, $s = 7$ or 15, with $d = 127$ and $L = 32$. Hence $\deg(g(z)) = 7$ or 15. The period of the sequence is very large, namely, $2^{127} - 1$, since the instances of $M(z)$ above are primitive polynomials. The range for the case of $s = 7$ is at most $[-3.417, 3.728]$ from (5.2). For $s = 15$, the range is at most $[-4.771, 4.997]$ also from (5.2). Notice that the bounds given by $\inf_{t>0} B(t)$ and $\sup_{t>0} B(t)$ become tight when the period is large enough.

*   *   *

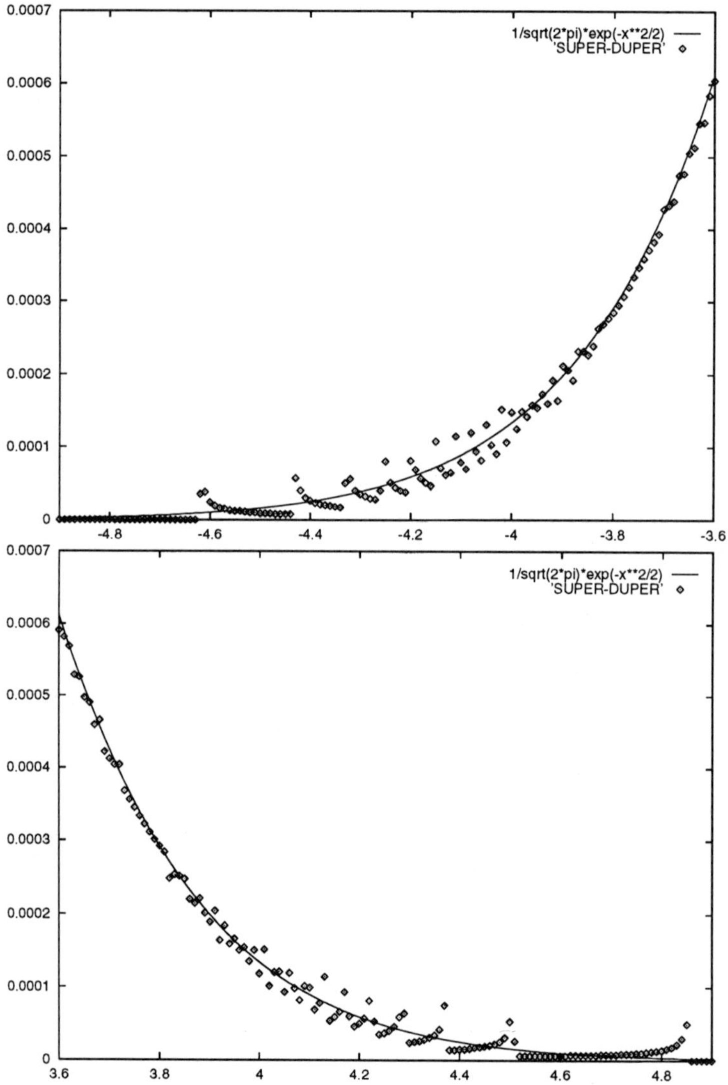

Figure 5.2 Tail distribution in the Box-Muller method with the combined $LS(2)$ sequence, a component of SUPER-DUPER, over the entire period.

Table 5.3 Frequencies of V_1 in the tail generated by the Box-Muller method with the combined $LS(2)$ sequences from SUPER-DUPER, over the entire period

Range	Observed O	Expected E	Deviation $(O-E)/\sqrt{E}$
$-3.94 : -3.95$	7451	7151	3.5
$-3.93 : -3.94$	6880	7438	-6.4
$-3.92 : -3.93$	8265	7735	6.0
$-3.91 : -3.92$	7026	8048	-11.3
$-3.90 : -3.91$	9117	8349	8.4
$-3.89 : -3.90$	8832	8718	1.2
$-3.88 : -3.89$	8238	9045	-8.4
$-3.87 : -3.88$	9959	9405	5.7
$-3.86 : -3.87$	9952	9775	1.7
$-3.85 : -3.86$	9757	10157	-3.9
$3.65 : 3.66$	21345	21517	-1.1
$3.66 : 3.67$	21053	20757	2.0
$3.67 : 3.68$	19716	20014	-2.1
$3.68 : 3.69$	20025	19284	5.3
$3.69 : 3.70$	18138	18622	-3.5
$3.70 : 3.71$	17716	17910	-1.4
$3.71 : 3.72$	17372	17265	0.8
$3.72 : 3.73$	17336	16625	5.5
$3.73 : 3.74$	15791	16020	-1.8
$3.74 : 3.75$	15295	15431	-1.0

Notes and Remarks

The OPSO and birthday spacings tests are from the stringent statistical tests proposed in Marsaglia [57]. The nearest pair test is due to Ripley [78]. All the other specific tests are from the standard statistical tests fully described in Knuth [40].

We can consider as many tests as there are algorithms for generation of non-uniform random variables, provided that the hypothesis H_0 holds. For example, Afflerbach and Wenzel [2] considered the sensitivity of linear congruential sequences to the Polar method, a variant of the Box-Muller method.

6

DERANDOMIZATION

Nothing in Nature is random ....
A thing appears random only through
the incompleteness of our knowledge.
– Spinoza, *Ethics* I

Gott würfelt nicht.
– Albert Einstein

Computer algorithms that involve randomness are often very useful because they can be not only easily implemented, but also are very efficient in performance [63]. Derandomization is a process that makes such algorithms deterministic. Some people consider that derandomization is an important process in algorithm design, mainly because modern computers are absolutely deterministic and thus are unable to produce true randomness. In fact, much effort has recently been devoted to derandomization of various randomized algorithms, and several general techniques have been devised for this purpose. In this chapter we will describe two important techniques, which are based on: low-discrepancy sequences and k-wise independent random variables.

6.1 LOW-DISCREPANCY SEQUENCES

The quasi-Monte Carlo method, which is the deterministic version of the Monte Carlo method, is very useful for numerical integration and global optimization problems. Monte Carlo methods assume (truly) random sequences, which cannot be generated on a computer, whereas quasi-Monte Carlo methods use low-discrepancy sequences instead, which are deterministically generated sequences, and are thus available on a computer. The theoretical advantage of using low-discrepancy sequences is that the convergence rate of numerical integration with these sequences outperforms that with random sequences.[1] In practice, however, the theoretical prediction is not necessarily correct. The main problem at present is how to construct low-discrepancy sequences that perform very well not only in theory but also in practice. One of main recent trends in this connection is the construction of low-discrepancy sequences based on finite fields [72]. By extending Sobol's net theory, Niederreiter [69] developed a principle for the construction of low-discrepancy sequences, which he calls the "digital construction principle." Subsequently, he [70] proposed a new concrete construction method using formal Laurent series over finite fields, which has recently been generalized by Tezuka [91]. The sequences generated by this new method, hereafter called *generalized Niederreiter sequences*, are important for two main reasons: (1) the theoretical performance of the sequences is the best currently known, and (2) the sequences can be easily implemented and quickly generated.

6.1.1 (t, k)-Sequences and (t, m, k)-Nets

Using dyadic boxes, Sobol' [84] developed the so-called net theory, which has recently been generalized to general boxes in arbitrary bases by Niederreiter [72]. We need the following definitions:

DEFINITION **6.1** A b-ary box is an interval of the form

$$E = \prod_{i=1}^{k} [a_i b^{-d_i}, (a_i + 1) b^{-d_i})$$

with integers $d_i \geq 0$ and integers $0 \leq a_i < b^{d_i}$ for $1 \leq i \leq k$. □

DEFINITION **6.2** Let $0 \leq t \leq m$ be an integer. A (t, m, k)-net is a point set of b^m points in $[0, 1]^k$ such that $\#(E; b^m) = b^t$ for every b-ary box E with $V(E) = b^{t-m}$.□

[1]We should notice that the domain of the algorithm is different, that is, Monte Carlo methods work for the L_2 space of functions, whereas quasi-Monte Carlo methods require something like a bounded total variation of functions, which is much smaller than the L_2 space.

Note that (t, m, k)-nets are always (u, m, k)-nets for all integers u with $t < u \leq m$.

DEFINITION **6.3** Let $t \geq 0$ be an integer. A (t, k)-sequence $X_0, X_1, ...,$ in base b is a sequence of points in $[0, 1]^k$ if for all integers $j \geq 0$ and $m > t$, the point set consisting of the $[X_n]_m$ with $jb^m \leq n < (j + 1)b^m$ is a (t, m, k)-net in base b, where $[X]_m$ denotes the coordinate-wise m-digit truncation in base b of X. □

Also note that (t, k)-sequences are always (u, k)-sequences for all integers $u > t$. By using the general technique described in Theorem 2.60, we obtain the following theorem, which gives an important relation between (t, k)-sequences and (t, m, k)-nets [72, Lemma 4.22].

Theorem 6.1 *If there exists a (t, k)-sequence in base b, then there exists a $(t, m, k + 1)$-net in base b for every $m \geq t$.*

Following Sobol' and Faure's results, Niederreiter obtained the following theorem [69] for an arbitrary integer base $b \geq 2$:

Theorem 6.2 *For any $N > 1$, the discrepancy of the first N points of a (t, k)-sequence in base b satisfies*

$$D_N^{(k)} \leq c(t, k, b)\frac{(\log N)^k}{N} + O\left(\frac{(\log N)^{k-1}}{N}\right), \tag{6.1}$$

where $c(t, k, b) \approx \frac{b^t}{k!}\left(\frac{b}{2 \log b}\right)^k$.□

This means that if t does not depend N, then the (t, k)-sequence becomes a low-discrepancy sequence. Note that a smaller value of t gives a lower discrepancy. The following result due to Niederreiter [72]:

Proposition 6.1 *A $(0, k)$-sequence in base b can exist only if $k \leq b$.*

Note that the net theory does not provide an explicit way of constructing low-discrepancy sequences, nor does it prove the existence of such sequences.

6.1.2 Generalized Niederreiter Sequences

Next, we describe Niederreiter's principle [69] for constructing (t, k)-sequences in base b. Let $k \geq 1$ and $b \geq 2$ and $B = \{0, 1, ..., b-1\}$. Then we choose
(i) a commutative ring R with identity and card $(R) = b$;
(ii) bijections $\psi_r : B \to R$ for $r = 1, 2, ...$, with $\psi_r(0) = 0$ for all sufficiently large r;
(iii) bijections $\lambda_{ij} : R \to B$ for $1 \leq i \leq k$ and $j = 1, 2, ...$, with $\lambda_{ij}(0) = 0$ for $1 \leq i \leq k$ and all sufficiently large j;
(iv) elements $c_{jr}^{(i)} \in R$ for $1 \leq i \leq k, 1 \leq j$, and $1 \leq r$.

Niederreiter's construction is described as follows:

DEFINITION **6.4** For $1 \leq i \leq k$ and $0 \leq n$, define the ith coordinate of the point P_n as

$$P_n^{(i)} = \sum_{j=1}^{\infty} x_{nj}^{(i)} b^{-j},$$

where

$$x_{nj}^{(i)} = \lambda_{ij} \left(\sum_{r=1}^{\infty} c_{jr}^{(i)} \psi_r(a_r(n)) \right),$$

for $1 \leq j$. Here $n = \sum_{r=1}^{\infty} a_r(n) b^{r-1}$ is the digit expansion of n in base b. □

We call the matrix $C^{(i)} = (c_{jr}^{(i)})_{1 \leq j,r}$ the generator matrix of the ith coordinate of (t, k)-sequences. Hereafter, we assume that b is a prime power. Then R in the construction principle becomes the finite field $GF(b)$. Let

$$\mathbf{c}_m^{(i)}(l) = (c_{m,1}^{(i)}, ..., c_{m,l}^{(i)}) \in GF(b)^l,$$

and let

$$C(d_1, ..., d_k; l) = \{\mathbf{c}_m^{(i)}(l) \mid 1 \leq m \leq d_i, 1 \leq i \leq k\}.$$

Define $\rho(C; l)$ as the maximal integer d such that $C(d_1, ..., d_k; l)$ is linearly independent over $GF(b)$ for all nonnegative integers $d_1, ..., d_k$ satisfying $\sum_{1 \leq i \leq k} d_i = d$. Then we have

Theorem 6.3 *Let b be a prime power and let $R = GF(b)$. If an integer $t \geq 0$ satisfies*

$$t \geq l - \rho(C; l)$$

for each integer $l > t$, then the sequence given in Definition 6.4 is a (t, k)-sequence in base b.

This result tells us that in order to construct low-discrepancy sequences, we need generator matrices, $C^{(i)}, 1 \leq i \leq k$, such that $\rho(C; l)$ is as large as possible for all $l > t$. So, we now describe how to construct such generator matrices so that we obtain low-discrepancy sequences. They are called generalized Niederreiter sequences [91]. The approach is based on the formal Laurent series expansions over finite fields. Denote $S(z) \in GF\{b, z\}$ by

$$S(z) = \sum_{j=w}^{\infty} a_j z^{-j},$$

where all $a_j \in GF(b)$ and w is an arbitrary integer. Hereafter, we use the following notations: $[S(z)]$ means the polynomial part of $S(z) \in GF\{b, z\}$ and $[S(z)]_{p(z)} \stackrel{\text{def}}{=} [S(z)] \pmod{p(z)}$ with $0 \leq \deg([S(z)]_{p(z)}) < \deg(p(z))$.

Let polynomials $p_1(z), ..., p_k(z) \in GF[b, z]$ be pairwise coprime and let $e_i = \deg(p_i) \geq 1$ for $1 \leq i \leq k$. For $m \geq 1, 1 \leq i \leq k$, and $j \geq 1$, consider the expansion

$$\frac{y_{im}(z)}{p_i(z)^j} = \sum_{r=w}^{\infty} a^{(i)}(j, m, r) z^{-r}, \tag{6.2}$$

by which the elements $a^{(i)}(j, m, r) \in GF(b)$ are determined. Here $w \leq 0$ may depend on i, j, m, and each $y_{im}(z)$ is a polynomial such that the residue polynomials $[y_{im}(z)]_{p_i(z)}, (j-1)e_i \leq m-1 < je_i$, are linearly independent over $GF(b)$ for any $j > 0$ and $1 \leq i \leq k$. Then define

$$c_{mr}^{(i)} = a^{(i)}(m_i + 1, m, r), \tag{6.3}$$

for $1 \leq i \leq k, m \geq 1$, and $r \geq 1$, where $m_i = [(m-1)/e_i]$.

Remark **1** The van der Corput sequence is equivalent to a generalized Niederreiter sequence such that $k = 1$; $b = 2$; all the mappings λ_{1j}, ψ_r are identity mappings; $p_1(z) = z$; and all $y_{1m}(z) = 1$. □

Remark **2** Sobol' sequences are equivalent to generalized Niederreiter sequences in base $b = 2$ such that all the mappings λ_{ij}, ψ_r are identity mappings; $p_1(z) = z$, and for $i = 2, ..., k$, $p_i(z)$ is the $(i-1)$th primitive polynomial in the list of all primitive polynomials sorted according to nondecreasing degrees; $y_{im}(z) = y_{ih}(z)$, where $h = |m-1|_{e_i} + 1$ and $\deg(y_{ih}) = e_i - h$. □

Remark **3** Faure sequences are equivalent to generalized Niederreiter sequences such that $b \geq k$ is prime; $p_i(z) = z - i + 1$ for $1 \leq i \leq k$; and all $y_{im}(z) = 1$. □

Remark 4 The original Niederreiter sequences [70] are equivalent to generalized Niederreiter sequences such that $y_{im}(z) = z^h g_{ij}(z)$; $p_i(z), 1 \le i \le k$ is the ith polynomial in the list of all monic irreducible polynomials sorted according to nondecreasing degrees, where $0 \le h < e_i$ and $j \ge 1$ satisfies $m - 1 = je_i + h$; and $g_{ij}(z)$ is a polynomial with $\gcd(g_{ij}, p_i) = 1$ for all j. □

The following lemma plays a key role in the proof of Theorem 6.4.

Lemma 6.1 *Let the elements $c_{mr}^{(i)} \in GF(b)$ be given by (6.3). Then for each integer $l > \sum_{i=1}^k (e_i - 1)$ and any integers $d_1, ..., d_k \ge 0$ with*

$$1 \le \sum_{i=1}^{k} d_i \le l - \sum_{i=1}^{k} (e_i - 1),$$

the vectors

$$\mathbf{c}_m^{(i)}(l) = (c_{m,1}^{(i)}, ..., c_{m,l}^{(i)}) \in GF(b)^l, \tag{6.4}$$

$1 \le m \le d_i, 1 \le i \le k$, are linearly independent over $GF(b)$.

Proof: Let $d_1, ..., d_k$ be as in the lemma and suppose that the vectors in (6.4) satisfy a linear dependence relation

$$\sum_{i=1}^{k} \sum_{m=1}^{d_i} f_m^{(i)} \mathbf{c}_m^{(i)}(l) = \mathbf{0} \in GF(b)^l$$

with all $f_m^{(i)} \in GF(b)$. This can be written as

$$\sum_{i=1}^{k} \sum_{m=1}^{d_i} f_m^{(i)} a^{(i)}(m_i + 1, m, r) = 0,$$

for $1 \le r \le l$, where $m_i = [(m-1)/e_i]$. Denote

$$y_{im}(z) = y_{im}^{(0)}(z) p_i(z)^{m_i+1} + y_{im}^{(1)}(z) p_i(z)^{m_i} + \cdots + y_{im}^{(m_i)}(z) p_i(z) + y_{im}^{(m_i+1)}(z),$$

where $y_{im}^{(0)}(z) = [y_{im}(z)/p_i(z)^{m_i+1}]$ and $y_{im}^{(q)}(z) = [y_{im}(z)/p_i(z)^q]_{p_i(z)}$ for $1 \le q \le m_i + 1$. Note that $y_{im}^{(q)}(z) = 0$ for $q > m_i + 1$. By considering only the fractional part, we have

$$\sum_{i=1}^{k} \sum_{q=1}^{q_i+1} \sum_{m=1}^{d_i} f_m^{(i)} \frac{y_{im}^{(q)}(z)}{p_i(z)^q} = \sum_{r=1}^{\infty} (\sum_{i=1}^{k} \sum_{m=1}^{d_i} f_m^{(i)} a^{(i)}(m_i + 1, m, r)) z^{-r} = 0,$$

where $q_i = [(d_i - 1)/e_i]$. Denote

$$f_{iq}(z) = \sum_{m=1}^{d_i} f_m^{(i)} y_{im}^{(q)}(z)$$

for $1 \le q \le q_i + 1, 1 \le i \le k$. We have $\deg(f_{iq}) < e_i$. By using the uniqueness of the partial fraction decomposition as in the proof of Theorem 4.49 in Niederreiter [72], we obtain $f_{iq}(z) = 0$ for all $1 \le q \le q_i + 1, 1 \le i \le k$. Thus, for $q = q_i + 1$,

$$f_{iq}(z) = \sum_{m=q_ie_i+1}^{d_i} f_m^{(i)} y_{im}^{(q)}(z) = 0.$$

Since $y_{im}^{(q_i+1)}(z), q_ie_i \le m - 1 < d_i (\le (q_i + 1)e_i)$, are linearly independent over $GF(b)$, we have $f_m^{(i)} = 0$ for $q_ie_i \le m - 1 < d_i$ and $1 \le i \le k$. By repeating the same argument for $q = q_i, ..., 1$, it follows that all $f_m^{(i)} = 0$. □

Thus we have the following theorem:

Theorem 6.4 *Let $t = \sum_{i=1}^{k}(e_i - 1)$, where $e_i = \deg(p_i)$, $1 \le i \le k$. For all $N > 1$, the discrepancy of the first N points of a generalized Niederreiter sequence in base b is given as*

$$D_N^{(k)} \le c(t, k, b)\frac{(\log N)^k}{N} + O\left(\frac{(\log N)^{k-1}}{N}\right), \tag{6.5}$$

where $c(t, k, b)$ is the same as that of (t, k)-sequences.

Proof: From the results on (t, k)-sequences in Theorems 6.1 and 6.3 combined with Lemma 6.1, the proof is complete. □

Recall that smaller t gives lower discrepancy for (t, k)-sequences. A direct consequence from the above theorem is that in order to get a $(0, k)$-sequence, b should be chosen to be larger than or equal to k. Remark 3 motivates the following definition:

DEFINITION **6.5** *Generalized Faure sequences are defined as $(0, k)$-sequences obtained from generalized Niederreiter sequences.*

In the case of $b < k$, the minimum value of t, denoted by $T_b(k)$, is obtained by choosing $p_1(z), ..., p_k(z)$ as the first k monic irreducible polynomials over $GF(b)$ in the order of nondecreasing degrees [72]. Hence, for such a choice of $p_1(z), ..., p_k(z)$, we have

$$T_b(k) = \sum_{i=1}^{k} (\deg(p_i) - 1).$$

The following theorem, due to Niederreiter [72], tells us how small $T_b(k)$ can be.

Theorem 6.5 *Let b be any prime power. If $1 \leq k \leq b$, then $T_b(k) = 0$; otherwise, i.e., for $k > b$,*

$$T_b(k) < k(\log_b k + \log_b \log_b k + 1).$$

Therefore, when $k \leq b$, we have

$$c(T_b(k), k, b) < \frac{1}{k!}(\frac{k}{\log 2k})^k,$$

which means that $c(T_b(k), k, b)$ approaches 0 at a super-exponential rate as k grows. However, we should note that the growth of k implies that b also goes to infinity. For a fixed b, if k grows, $c(T_b(k), k, b) \approx b^{k \log \log k}$, which is slightly better than the constant factor $\approx b^{k \log k}$ for Halton sequences. The case $b = 2$ is practically the most important. Thus, Remark 2 motivates the following:

DEFINITION **6.6** *Generalized Sobol' sequences are defined as generalized Niederreiter sequences in base $b = 2$ such that all the mappings λ_{ij}, ψ_r are identity mappings; $p_i(z), 1 \leq i \leq k$ is the ith polynomial in the list of all monic irreducible polynomials sorted according to nondecreasing degrees.*

Efficient Implementation and Practical Performance

Antonov and Saleev [6] proposed an efficient generation of (t, k)-sequences in base $b = 2$ by using the Gray code. For brevity, we consider the one-dimensional case $k = 1$ with base $b = 2$ and the mappings λ_{ij} and ψ_r being the identity. Denote a sequence by $X_n, n = 0, 1, 2, ..., 2^m - 1$, where m is large enough for practical purposes, say $m = 32$. From the definition, X_n is obtained by the multiplication of the $m \times m$ generator matrix $C = (c_{ij})$ by the index vector, that is, $C\mathbf{n}$, where $\mathbf{n} = {}^t(n_1, ..., n_m)$ for an integer n whose binary representation is $n = n_m 2^{m-1} + \cdots + n_2 2 + n_1$. Consider the Gray code $\mathbf{g} = {}^t(g_1, ..., g_m)$ such that $g_j = n_j + n_{j+1} \pmod 2$ for $j = 1, ..., m$, where we assume $n_{m+1} = 0$. Antonov and Saleev showed that

$$C\mathbf{g}_n, \quad n = 0, 1, ..., 2^m - 1,$$

also constitute the first 2^m points of a (t, k)-sequence. The proof is easy because the matrix corresponding to the Gray code is nonsingular.

This fact has a great practical merit. Indeed, consider the difference between adjacent elements in the sequence,

$$C\mathbf{g}_n \text{ XOR } C\mathbf{g}_{n+1} = C(\mathbf{g}_n \text{ XOR } \mathbf{g}_{n+1}).$$

Since the vector, $\mathbf{g}_n$ XOR $\mathbf{g}_{n+1}$, is always a binary vector of weight 1, by tracing the column vector of C corresponding to the nonzero component of that vector, we can easily compute the next element in the sequence from the present element.

Next, we make the above idea applicable to (t, k)-sequences in a general base b by using the b-ary Gray code, which is defined as follows: Define an $m \times m$ matrix

$$G = \begin{pmatrix} 1 & -1 & & & \\ & 1 & -1 & & \\ & & \cdots & & \\ & & & 1 & -1 \\ & & & & 1 \end{pmatrix} \pmod b.$$

The b-ary Gray code ${}^t(g_1^{(n)}, ..., g_m^{(n)}), n = 0, 1, 2, ...,$ is defined by $G\,{}^t(n_1, ..., n_m)$, where ${}^t(n_1, ..., n_m)$ is a vector of length m obtained by expanding an integer n in base b, that is, $n = n_m b^{m-1} + \cdots + n_2 b + n_1$. We then easily obtain the next theorem:

Theorem 6.6 *Let $\mathbf{e}_j$ be the jth unit vector of length m for $j = 1, 2, ..., m$. Then*

$$(g_1^{(n+1)}, ..., g_m^{(n+1)}) = (g_1^{(n)}, ..., g_m^{(n)}) \text{ ADD}_b\ \mathbf{e}_j,$$

where ADD_b means the coordinate-wise modulo b addition and j is the smallest index with $n_j \neq b - 1$ in $(n_1, ..., n_m)$.

It is easy to see that this result leads us to the b-ary generalization of the above Antonov-Saleev implementation. See a sample C program for generalized Faure sequences, given in Figures A.3 of Appendix A.

Now, we employ the L_2-discrepancy to compare the performance of generalized Faure sequences with those of the original Faure sequences, Halton sequences, and Sobol' sequences. By using the formula (2.8), we computed the exact values of the L_2-discrepancy in 5, 10, 15, and 20 dimensions for the original Faure, generalized Faure, Sobol', Halton, and generalized Halton sequences, due to Warnock, as given in Section 2.3. For every coordinate of the Sobol' sequence, we set all the starting values to be 1 for the recurrence of direction numbers. In other words, we set $y_{ih}(z) = [p_i(z)/z^h]$ for all $1 \le h \le e_i$ and $1 \le i \le k$ in Remark 2. In 15 and 20 dimensions, all values of the L_2-discrepancy except those for generalized Faure sequences are worse than those for random sequences. This result agrees with those obtained in earlier studies [15, 62, 106].

Polynomial Arithmetic Analogue of Halton Sequences

In the 1960's, there were two important classes of low-discrepancy point sets and sequences: Good lattice points and Halton sequences. An analogous version of the first class with respect to polynomial arithmetic over finite fields has recently been defined in general and analyzed in detail by Niederreiter [72], who showed that the point sets were an example of (t, m, k)-nets satisfying his construction principle. Now, we will show that generalized Niederreiter sequences include as their special subclass the polynomial arithmetic analogue of Halton sequences. Thus, for this analogous version of Halton sequences, we can obtain asymptotically much smaller constant term in the upperbound of discrepancy than for the original Halton sequences.

First, we define the radical inverse function with respect to polynomial arithmetic over the finite field $GF(b)$, where b is a prime power.

DEFINITION **6.7** Let n be a nonnegative integer whose digit expansion in prime power base b is $n = a_{m+1}(n)b^m + \cdots + a_2(n)b + a_1(n)$, where $m = \log_b n$. Denote $v_n(z) = n_m z^m + \cdots + n_1 z + n_0$, where $n_i = \psi_{i+1}(a_{i+1}(n)), i = 0, 1, ..., m$. Let $p(z)$ be an arbitrary non-constant polynomial over $GF(b)$. Then $v_n(z)$ can be represented in terms of $p(z)$ as follows: Let $e = \deg(p)$ and $s = [m/e]$.

$$v_n(z) = r_s(z)p(z)^s + \cdots + r_1(z)p(z) + r_0(z),$$

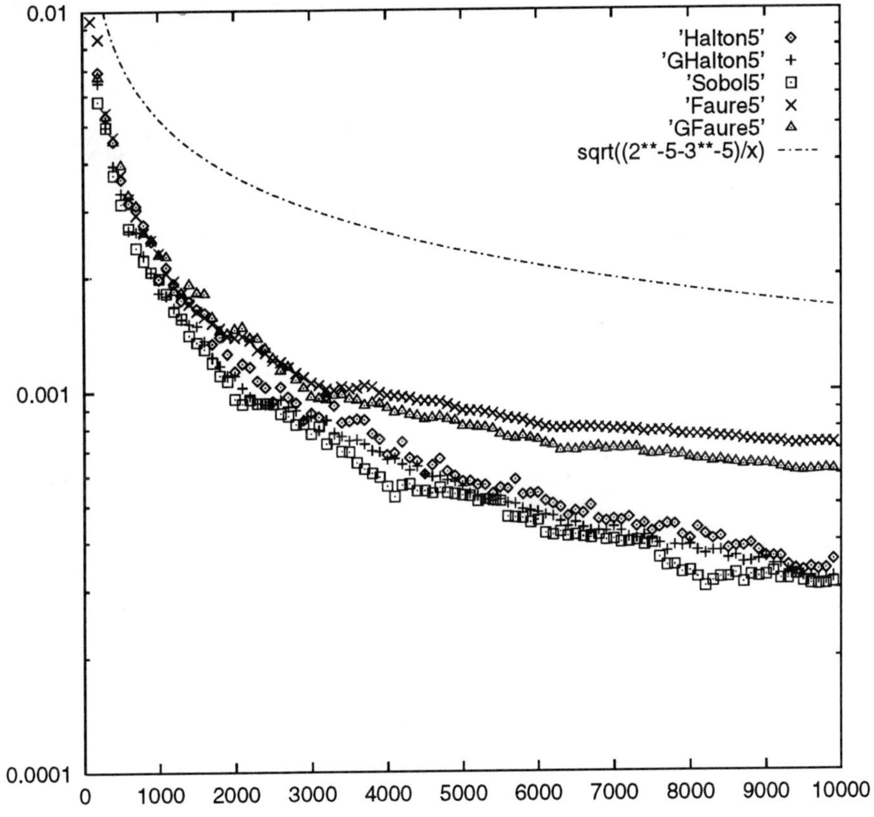

Figure 6.1 L_2-discrepancy in five dimensions for the first 10000 points, $X_n, n = 1, ..., 10000$, from five types of low-discrepancy sequence.

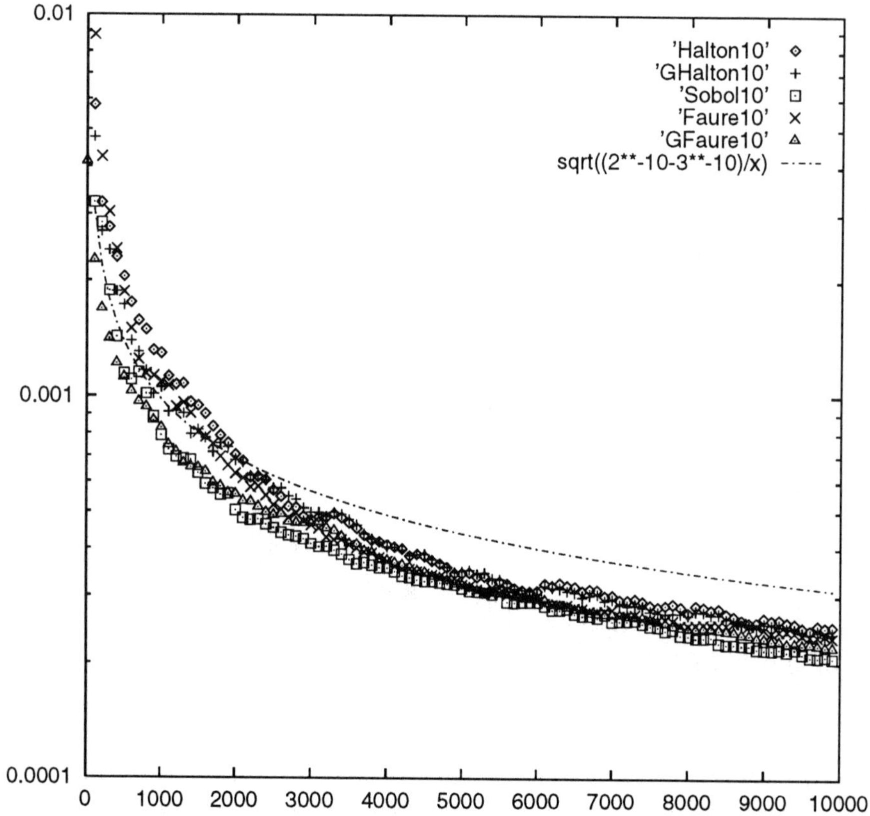

Figure 6.2 L_2-discrepancy in 10 dimensions for the first 10000 points, $X_n, n = 1, ..., 10000$, from five types of low-discrepancy sequence.

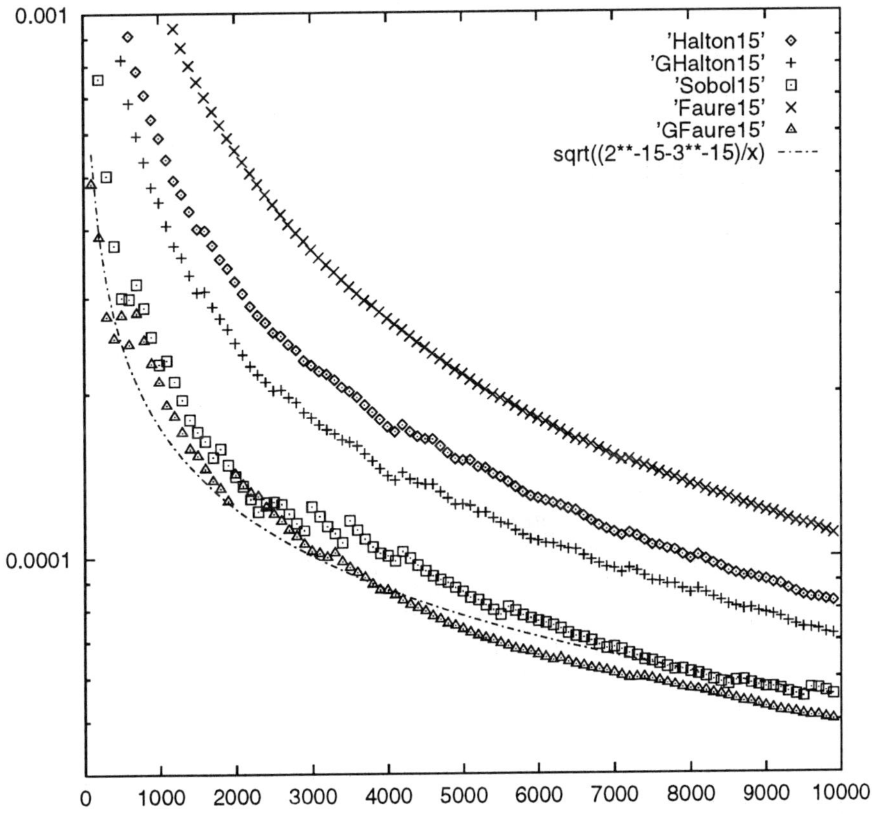

Figure 6.3 L_2-discrepancy in 15 dimensions for the first 10000 points, $X_n, n = 1, ..., 10000$, from five types of low-discrepancy sequence.

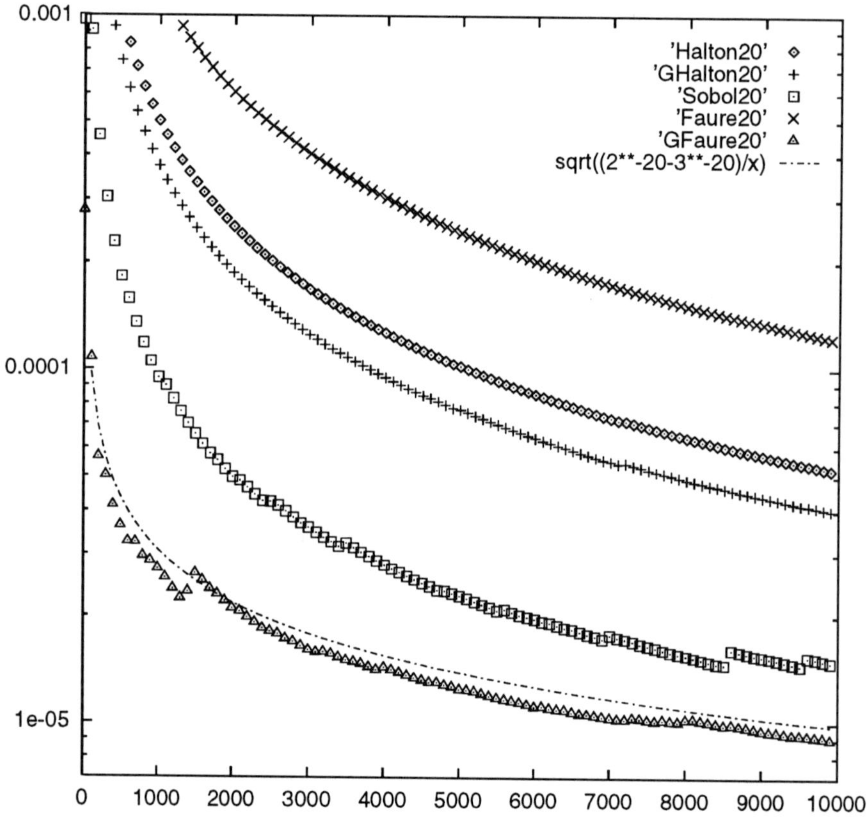

Figure 6.4 L_2-discrepancy in 20 dimensions for the first 10000 points, $X_n, n = 1, ..., 10000$, from five types of low-discrepancy sequence.

where $r_i(z) = [v_n(z)/p(z)^i]_{p(z)}, 0 \le i \le s$. Note that $e > \deg(r_i)$ for all $0 \le i \le s$. We define the radical inverse function $\phi_{p(z)} : GF[b,z] \to GF\{b,z\}$ as follows:

$$\phi_{p(z)}(v_n(z)) = \frac{r_0(z)}{p(z)} + \cdots + \frac{r_{s-1}(z)}{p(z)^s} + \frac{r_s(z)}{p(z)^{s+1}}.$$

□

We give two examples:

EXAMPLE **6.1** Take $p(z) = z \in GF[2,z]$, and let ψ_r be the identity mapping. Then, for $n = 0, 1, ..., 7$,

$$\begin{array}{ll} v_0(z) = 0, & \phi_z(v_0(z)) = 0, \\ v_1(z) = 1, & \phi_z(v_1(z)) = z^{-1}, \\ v_2(z) = z, & \phi_z(v_2(z)) = z^{-2}, \\ v_3(z) = z+1, & \phi_z(v_3(z)) = z^{-1} + z^{-2}, \\ v_4(z) = z^2, & \phi_z(v_4(z)) = z^{-3}, \\ v_5(z) = z^2+1, & \phi_z(v_5(z)) = z^{-1} + z^{-3}, \\ v_6(z) = z^2+z, & \phi_z(v_6(z)) = z^{-2} + z^{-3}, \\ v_7(z) = z^2+z+1, & \phi_z(v_7(z)) = z^{-1} + z^{-2} + z^{-3}. \end{array}$$

□

EXAMPLE **6.2** Take $p(z) = z+1 \in GF[2,z]$, and let ψ_r be the identity mapping. Then, for $n = 0, 1, ..., 7$,

$$\begin{aligned} \phi_{z+1}(v_0(z)) &= 0, \\ \phi_{z+1}(v_1(z)) &= \frac{1}{z+1} = z^{-1} + z^{-2} + \cdots, \\ \phi_{z+1}(v_2(z)) &= \frac{1}{z+1} + \frac{1}{(z+1)^2} = z^{-1} + z^{-3} + \cdots, \\ \phi_{z+1}(v_3(z)) &= \frac{1}{(z+1)^2} = z^{-2} + z^{-4} + \cdots, \\ \phi_{z+1}(v_4(z)) &= \frac{1}{z+1} + \frac{1}{(z+1)^3} = z^{-1} + z^{-2} + z^{-5} + z^{-6} + \cdots, \\ \phi_{z+1}(v_5(z)) &= \frac{1}{(z+1)^3} = z^{-3} + z^{-4} + z^{-7} + z^{-8} + \cdots, \\ \phi_{z+1}(v_6(z)) &= \frac{1}{(z+1)^2} + \frac{1}{(z+1)^3} = z^{-2} + z^{-3} + z^{-6} + z^{-7} + \cdots, \\ \phi_{z+1}(v_7(z)) &= \frac{1}{z+1} + \frac{1}{(z+1)^2} + \frac{1}{(z+1)^3} = z^{-1} + z^{-4} + z^{-5} + z^{-8} + \cdots, \end{aligned}$$

□

We now define the polynomial arithmetic version of Halton sequences.

DEFINITION **6.8** A Halton sequence with respect to polynomial arithmetic over finite fields is defined as follows: Let X_n be a k-dimensional vector in $[0,1)^k$ and let $p_1(z), ..., p_k(z)$ be pairwise coprime polynomials over $GF(b)$. Then, for $n = 0, 1, ...,$

$$X_n = (\sigma_1(\phi_{p_1(z)}(v_n(z))), ..., \sigma_k(\phi_{p_k(z)}(v_n(z))),$$

where each $\sigma_i, i = 1, ..., k$, is a mapping from $GF\{b, z\}$ to the real field defined by $\sigma_i(\sum_{j=w}^{\infty} a_j z^{-j}) = \sum_{j=w}^{\infty} \lambda_{ij}(a_j) b^{-j}$. □

EXAMPLE **6.3** Following the previous examples, consider the case $k = 2$ and $GF(2)$. Let $X_n = (\sigma_1(\phi_z(v_n(z))), \sigma_2(\phi_{z+1}(v_n(z))))$, and let λ_{ij} be the identity mapping. Then we have $(0,0), (\frac{1}{2}, 1), (\frac{1}{4}, \frac{2}{3}), (\frac{3}{4}, \frac{1}{3}), (\frac{1}{8}, \frac{4}{5}), (\frac{5}{8}, \frac{1}{5}), (\frac{3}{8}, \frac{2}{5}), (\frac{7}{8}, \frac{3}{5})$ for $n = 0, 1, ..., 7$. □

From Definition 6.7, we observe the following property of our analogue of a radical inverse function:

Property **6.1** *Let $v_n(z) = n_m z^m + \cdots + n_1 z + n_0$. We have*

$$\phi_{p(z)}(v_n(z)) = \sum_{j=0}^{m} n_j \phi_{p(z)}(z^j).$$

□

Denote for $m \geq 0$

$$\phi_{p(z)}(z^m) = \frac{r_0^{(m)}(z)}{p(z)} + \cdots + \frac{r_{s-1}^{(m)}(z)}{p(z)^s} + \frac{r_s^{(m)}(z)}{p(z)^{s+1}},$$

where $s = [m/e]$, and denote $\phi_{p(z)}(v_n(z)) = a_1 z^{-1} + a_2 z^{-2} + \cdots$. Then, it follows from the above property that $\mathbf{a}(l) = {}^t(a_1, a_2, ..., a_l)$ is a linear transformation of $\mathbf{n}(l) = {}^t(n_0, n_1, ..., n_{l-1})$, that is, $\mathbf{a}(l) = C(l)\mathbf{n}(l)$, where $C(l)$ is an $l \times l$ generator matrix $(c_{ij}; 1 \leq i, j \leq l)$ over $GF(b)$. Define the $r \times r$ Hankel matrix $H_r(S)$

associated with a formal Laurent series $S(z) = \sum_{i=1}^{\infty} x_i z^{-i}$ in the following way:

$$H_r(S) = \begin{pmatrix} x_1 & x_2 & x_3 & \cdots & x_r \\ x_2 & x_3 & x_4 & \cdots & x_{r+1} \\ & \cdots & \cdots & \cdots & \\ & \cdots & \cdots & \cdots & \\ x_{r-1} & x_r & x_{r+1} & \cdots & x_{2r-2} \\ x_r & x_{r+1} & x_{r+2} & \cdots & x_{2r-1} \end{pmatrix}.$$

The next lemma plays a key role in proving our main result, Theorem 6.7.

Lemma 6.2 *$C(l)$ is a symmetric matrix for every $l \geq 1$.*

Proof: We need some preparation before giving the main part of the proof. For every integer $i \geq 1$ and $l \geq i$, the elements of the ith column vector of $C(l)$ form a linear feedback shift register sequence whose characteristic polynomial is $p(z)^q$, where $q = [(i-1)/e] + 1$. First, we show that this is also true for each row vector of $C(l)$. For a given $m \geq e$, let $s = \lceil m/e \rceil$, and denote $p(z)^s = d_{es} z^{es} + d_{es-1} z^{es-1} + \cdots + d_0$ with $d_{es} \neq 0$. Then, for $0 \leq q < s$, we have $z^{m-es} p(z)^s = 0 \pmod{p(z)^{q+1}}$, i.e., $d_{es} z^m = -(d_{es-1} z^{m-1} + \cdots + d_0 z^{m-es}) \pmod{p(z)^{q+1}}$. Let $h = m - es$, then for $0 \leq q < s$,

$$\begin{aligned} r_q^{(m)}(z) &= [z^m / p(z)^q]_{p(z)} \\ &= -\sum_{j=0}^{es-1} d_{es}^{-1} d_j [z^{h+j} / p(z)^q]_{p(z)}. \end{aligned}$$

Thus, we obtain

$$\begin{aligned} -\sum_{j=0}^{es-1} d_{es}^{-1} d_j \phi_{p(z)}(z^{m-es+j}) &= -\sum_{j=0}^{es-1} d_{es}^{-1} d_j \sum_{q=0}^{s} \frac{[z^{h+j}/p(z)^q]_{p(z)}}{p(z)^{q+1}} \\ &= \sum_{q=0}^{s-1} \frac{r_q^{(m)}(z)}{p(z)^{q+1}} + \frac{\epsilon(z)}{p(z)^{s+1}} \\ &= \phi_{p(z)}(z^m) + \frac{\epsilon(z) - r_s^{(m)}(z)}{p(z)^{s+1}}, \end{aligned}$$

where $\epsilon(z) = -\sum_{j=0}^{es-1} d_{es}^{-1} d_j [z^{h+j}/p(z)^s]_{p(z)}$. Since $\nu((\epsilon(z) - r_s^{(m)}(z))/p(z)^{s+1}) < -es$, the characteristic polynomial of the ith row vector of $C(l)$ is $p(z)^q$, where $q = [(i-1)/e] + 1$ for $i \geq 1$.

Now we show, by induction, the symmetry of $C(se)$ for every $s \geq 1$. It is easy to see that $C(e)$ is symmetric because it is the $e \times e$ Hankel matrix associated with $r_0^{(0)}(z)/p(z)$. Assume that $C(se)$ is symmetric for some $s \geq 1$. From the above preparation, we have $c_{ij} = c_{ji}$, for $1 \leq i \leq se$ and $se < j \leq (s+1)e$. The submatrix $(c_{ij}; se < i, j \leq (s+1)e)$ is also symmetric, because it is equal to the sum of the following two $e \times e$ symmetric matrices: one corresponding to $r_s^{(m)}(z)/p(z)^{s+1}$, which is symmetric due to Lemma 6.3, given below, and the other corresponding to $\sum_{q=0}^{s-1} r_q^{(m)}(z)/p(z)^{q+1}$, which is also symmetric from the assumption. Thus, the proof is complete. □

Lemma 6.3 *Assume that* $p(z) \in GF[b, z]$ *with* $e = \deg(p) > 0$. *Let* $z^m = R_1^{(m)}(z)p(z)^s + R_0^{(m)}(z)$ *for* $se \leq m < (s+1)e$, *where* $R_1^{(m)}(z) = [z^m/p(z)^s]$ *and* $R_0^{(m)}(z) = [z^m]_{p(z)^s}$. *Then for* $1 \leq i, j \leq e$, *the coefficient of* z^{-se-j} *in the formal Laurent series* $R_1^{(se-1+i)}(z)/p(z)^{s+1}$ *is identical with the* $(se+i, se+j)$ *element of* $H_{(s+1)e}(z^{se}/p(z)^{2s+1})$.

Proof: This follows from the facts that $z^m/p(z)^{2s+1} = R_1^{(m)}(z)/p(z)^{s+1} + R_0^{(m)}(z)/p(z)^{2s+1}$ for $se \leq m < (s+1)e$, and that $\nu(R_0^{(m)}(z)/p(z)^{2s+1}) < -(s+1)e$. □

EXAMPLE **6.4** Following Example 6.2, consider the case $b = 2$, i.e., $GF(2)$. The matrix $C(8)$ for the sequence $\phi_{z+1}(v_n(z))$, $n = 0, 1, ..., 2^8 - 1$, is given by

$$C(8) = \begin{pmatrix} 1 & 1 & 1 & 1 & 1 & 1 & 1 & 1 \\ 1 & 0 & 1 & 0 & 1 & 0 & 1 & 0 \\ 1 & 1 & 0 & 0 & 1 & 1 & 0 & 0 \\ 1 & 0 & 0 & 0 & 1 & 0 & 0 & 0 \\ 1 & 1 & 1 & 1 & 0 & 0 & 0 & 0 \\ 1 & 0 & 1 & 0 & 0 & 0 & 0 & 0 \\ 1 & 1 & 0 & 0 & 0 & 0 & 0 & 0 \\ 1 & 0 & 0 & 0 & 0 & 0 & 0 & 0 \end{pmatrix}.$$

The matrix is symmetric. □

On the basis of the above lemma, we obtain the main result:

Theorem 6.7 *The polynomial arithmetic version of Halton sequences constitutes a class of low-discrepancy sequences whose discrepancy is given in Theorem 6.4.*

Proof: Denote

$$\phi_{p_i(z)}(z^m) = \frac{r_0^{(i,m)}(z)}{p_i(z)} + \cdots + \frac{r_{s-1}^{(i,m)}(z)}{p_i(z)^s} + \frac{r_s^{(i,m)}(z)}{p_i(z)^{s+1}},$$

for $1 \leq i \leq k$. From Lemma 6.2, we have

$$c_{mr}^{(i)} = c_{rm}^{(i)} = \sum_{q=1}^{m_i+1} a^{(i)}(q,m,r),$$

where $r_{j-1}^{(i,m-1)}(z)/p_i(z)^j = \sum_{r=w}^{\infty} a^{(i)}(j,m,r)z^{-r}$. Since $\deg(r_s^{(i,m-1)}) = m - 1 - se_i$, it follows that $r_s^{(i,m-1)}(z), se_i \leq m-1 < (s+1)e_i$, are linearly independent over $GF(b)$ for all $1 \leq i \leq k$. Therefore, our analogous Halton sequence, X_n, $n = 0, 1, ...,$ is equivalent to a generalized Niederreiter sequence with $y_{im}(z) = \sum_{j=0}^{s} r_j^{(i,m-1)}(z)p_i(z)^{s-j}$ for $m \geq 1$ and $1 \leq i \leq k$.

From Theorem 6.4 and the fact that for a fixed b, $c(b,t,k)$ can be made depending solely on k by choosing appropriate $p_i(z)$'s, the proof is complete. □

We have the following corollaries [98]:

Corollary 6.1 *If $b \geq k$ is a prime power, $(0,k)$-sequences in base b are obtained from our analogous Halton sequences by putting $p_i(z) = z - b_i$ for $1 \leq i \leq k$, where $b_1, ..., b_k$ are distinct elements in $GF(b)$.*

Corollary 6.2 *For $i = 1, ..., k$, the generator matrix $C^{(i)}$ of $(0,k)$-sequences in a prime base $b \geq k$ from our analogous Halton sequences is written as*

$$C^{(i)} = P^{i-1}\,({}^tP)^{i-1},$$

where P is Pascal's matrix[2].

[2] Pascal's matrix is a matrix whose (i,j) element is equal to $\binom{i-1}{j-1}$.

Proof: From the definition, by putting $p_i(z) = z - i + 1$ for $1 \leq i \leq k$, we denote

$$z^m = r_0 + \cdots + r_{m-1}(z-i)^{m-1} + r_m(z-i)^m.$$

By applying the binomial theorem to $z^m = (z - i + i)^m$, we get

$$r_q = \binom{m}{q} i^{m-q}$$

for $0 \leq q \leq m$. Note that $r_m = 1$. We also have

$$\frac{1}{z-i} = \sum_{j=0}^{\infty} \frac{i^j}{z^{j+1}}.$$

By repeating the multiplication, we obtain

$$\frac{1}{(z-i)^{q+1}} = \sum_{j=q}^{\infty} \frac{\binom{j}{q}}{z^{j+1}} i^{j-q}.$$

From the definition of

$$\phi_{z-i}(z^m) = \frac{r_0}{z-i} + \cdots + \frac{r_{m-1}}{(z-i)^m} + \frac{r_m}{(z-i)^{m+1}},$$

we determine that the $(m+1, j+1)$ element of the generator matrix is equal to the coefficient of the term z^{-j-1} in the formal Laurent series expansion of $\phi_{z-i}(z^m)$, that is,

$$\sum_{q=0}^{\min(m,j)} \binom{m}{q} \binom{j}{q} i^{m+j-2q}.$$

This completes the proof. □

Note that the generator matrix of the ith coordinate of Faure sequences is $({}^tP)^{i-1}$.

6.1.3 Discrepancy and Numerical Integration

First, we define the total variation of a function on $[0,1]^k$, which reflects the degree of regularity of the function.

DEFINITION **6.9** The k-dimensional total variation of a function f on $[0,1]^k$ in the sense of Vitali is defined by

$$V^{(k)}(f) = \sup_{\pi_1,\ldots,\pi_k} \sum_{j_1=1}^{n_1} \cdots \sum_{j_k=1}^{n_k} |\delta_{j_1}^{(1)} \cdots \delta_{j_k}^{(k)} f(x_1,\ldots,x_k)|,$$

where we define

$$\begin{aligned}\delta_{j_i}^{(i)} f(x_1, ..., x_k) &= f(x_1, ..., x_{i-1}, t_{i,j_i}, x_{i+1}, ..., x_k) \\ &\quad - f(x_1, ..., x_{i-1}, t_{i,j_i-1}, x_{i+1}, ..., x_k),\end{aligned}$$

and $0 = t_{i0} < t_{i1} < \cdots < t_{in_i} = 1$ $(i = 1, ..., k)$ are k arbitrary partitions $\pi_1, ..., \pi_k$ of each dimension of $[0,1]^k$. The function f is said to be of bounded variation in the sense of Vitali, if $V^{(k)}(f)$ is finite. □

DEFINITION **6.10** For $1 \le h \le k$ and $1 \le i_1 < \cdots < i_h \le k$, let $f^{(i_1,...,i_h)}(x_1, ..., x_k)$ be the restriction of $f(x_1, ..., x_k)$ to the h-dimensional face $\{(u_1, ..., u_k) \in [0,1]^k \mid u_j = 1 \text{ for } j \neq i_1, ..., i_h\}$. Then we define

$$V(f) = \sum_{h=1}^{k} \sum_{1 \le i_1 < \cdots < i_h \le k} V^{(h)}(f^{(i_1,...,i_h)})$$

as the variation of the function f in the sense of Hardy and Krause. The function f is said to be of bounded variation in the sense of Hardy and Krause, if $V(f)$ is finite. □

The use of low-discrepancy sequences for numerical integration is justified by the following Koksma-Hlawka theorem, which tells us the importance of the discrepancy for error estimates of numerical integration:

Theorem 6.8 *If a function f on $[0,1]^k$ is of bounded variation in the sense of Hardy and Krause, and $\mathbf{x}_1, \mathbf{x}_2, ...,$ is a sequence of points in $[0,1)^k$, then*

$$E_N(f) \stackrel{\text{def}}{=} \left| \int_{[0,1]^k} f(\mathbf{x}) d\mathbf{x} - \frac{1}{N} \sum_{i=0}^{N-1} f(\mathbf{x}_i) \right| \le V(f) D_N^{(k)}.$$

Note that this result gives deterministic error-bounds for numerical integration instead of probabilistic error-bounds for Monte Carlo integration.

For the L_2-discrepancy $T_N^{(k)}$, Woźniakowski [109] recently obtained the following:

Theorem 6.9 *Let C_k be the class of real continuous functions defined on $[0,1]^k$ equipped with the classical Wiener sheet measure w (that is, Gaussian with mean* *zero and covariance kernel*

$$R(\mathbf{s},\mathbf{t}) \stackrel{\text{def}}{=} \int_{C_k} f(\mathbf{s})f(\mathbf{t})\, w(df) = \min(\mathbf{s},\mathbf{t}) \stackrel{\text{def}}{=} \prod_{j=1}^{k} \min(s_j,t_j)$$

for any vectors $\mathbf{s} = (s_1, ..., s_k)$ *and* $\mathbf{t} = (t_1, ..., t_k)$ *in* $[0,1]^k$). *Then, for a given set of points* $\mathbf{x}_i = (x_{i1}, ..., x_{ik}), i = 0, 1, ..., N-1$, *in* $[0,1]^k$, *we have*

$$\int_{C_k} (E_N(f))^2\, w(df) = (\bar{T}_N^{(k)})^2,$$

where $\bar{T}_N^{(k)}$ *is the* L_2*-discrepancy of the point set* $\bar{\mathbf{x}}_i = (1-x_{i1}, ..., 1-x_{ik}), i = 0, 1, ..., N-1$.

The result means that on average the integration error is dependent only on the discrepancy, not on the integrand f at all, and that low-discrepancy sequences are useful in numerical integration.

6.1.4 Dispersion and Global Optimization

Another important application of low-discrepancy sequences is global optimization. Consider the following type of continuous optimization problem. Let X be a separable topological space and assume that we want to find a global optimum of a real-valued function f on X. Define

$$m(f) \stackrel{\text{def}}{=} \sup_{x \in X} f(x),$$

where we assume that the function f is bounded from above. The well-known random search procedure takes a sequence of independent λ-distributed random samples $P = \{x_0, x_1, ... \in X\}$, where λ is a probability measure on X, to get the estimate

$$m(f) \approx m_N(f;P) \stackrel{\text{def}}{=} \max_{0 \le n \le N-1} f(x_n).$$

Note that $m_N(f;P), N = 1, 2, ..,$ is a nondecreasing sequence. We have the following result concerning its convergence [72]:

Theorem 6.10 *If a function* f *is continuous on* X *and if* λ *is such that* $\lambda(A) > 0$ *for every nonempty open subset* A *of* X, *then*

$$\lim_{N \to \infty} m_N(f;P) = m(f),$$

with probability 1.

The deterministic version of random search, called quasi-random search, is a typical example of the quasi-Monte Carlo method. Here we use a deterministic sequence of points in X. It is clear that

$$\lim_{N \to \infty} m_N(f; P) = m(f),$$

if a function f is continuous on X and the sequence is dense in X. For the case of a finite sequence of points, we have an important notion, *dispersion*, which is very useful for estimating the error of quasi-random search.

DEFINITION **6.11** Let *dist* be the metric defined in X satisfying $\sup_{x,y \in X} dist(x, y) < \infty$. For any point sequence P in X, the dispersion of the first N points of P is defined as

$$d_N(P; X) = \sup_{x \in X} \min_{1 \leq n \leq N} dist(x, x_n).$$

The modulus of continuity of f is defined as

$$\omega(f; t) = \sup_{\substack{x, y \in X \\ dist(x, y) \leq t}} |f(x) - f(y)|$$

for all $t \geq 0$. □

Note that for a given function f, the modulus of continuity $\omega(f; t)$ is non-decreasing as t grows. We thus have the following relation [72] between the dispersion and the error of quasi-random search.

Theorem 6.11 *For any point sequence P, if its first N points have the dispersion $d_N(P; X)$, then*

$$m(f) - m_N(f; P) \leq \omega(f; d_N).$$

This result ensures that a point set with smaller dispersion gives better solutions for quasi-random search.

Hereafter, for simplicity we assume that $X = [0, 1]^k$ and that *dist* is the maximum (or L_∞) metric. In this case, there are upper and lower bounds for the dispersion.

Theorem 6.12 *For any point sequence P in $[0,1]^k$, its first N points satisfy*

$$\frac{1}{2[N^{1/k}]} \leq d_N(P) \leq (D_N^{(k)})^{1/k},$$

where $d_N(P)$ denotes $d_N(P;[0,1]^k)$ with the maximum metric.

The first inequality is known to be tight. The second inequality shows a connection between dispersion and discrepancy, which means that low-discrepancy sequences can be used as low-dispersion sequences. Since there exist low-discrepancy sequences with $D_N^{(k)} = O(N^{-1}(\log N)^k)$ in k dimensions, we obtain $d_N(P) = O(N^{-1/k}\log N)$, where the constant term is dependent only on the dimension k. Niederreiter [72] analyzed the constant term for several low-discrepancy sequences:

Theorem 6.13 *For all $N \geq 1$, the first N points of the Halton sequence with pairwise coprime bases $b_1, ..., b_k$ satisfy*

$$d_N(P) < \frac{\max_{1\leq i\leq k} b_i}{N^{1/k}},$$

and the Hammersley N-point set with pairwise relatively prime bases $b_1, ..., b_{k-1}$ satisfies

$$d_N(P) < \frac{1 + \max_{1\leq i\leq k-1} b_i}{N^{1/k}}.$$

Theorem 6.14 *For all $N \geq 1$, the first N points of any (t,k)-sequence in base b satisfy*

$$d_N(P) < \frac{b^{(k+t)/k}}{N^{1/k}},$$

and any (t,m,k)-net in base b satisfies

$$d_N(P) \leq \frac{b^{(k+t-1)/k}}{N^{1/k}},$$

where $N = b^m$.

Note that for Halton sequences whose bases $b_1, ..., b_k$ are the first k small primes, the constant term becomes $\max_{1\leq i\leq k} b_i \approx k\log k$. On the other hand, for (t, k)-sequences in base b such that $p_1(z), ..., p_k(z)$ are chosen to be monic irreducible polynomials with the k smallest degrees, we have

$$b^{(k+T_b(k))/k} \approx \begin{cases} b & \text{if } k \leq b, \\ k \log_b k & \text{otherwise.} \end{cases}$$

Thus, from the theoretical viewpoint, the performance for the two types of point set is almost comparable, provided that b is fixed and $k > b$; otherwise, $(0, k)$-sequences are superior.

6.2 K-WISE INDEPENDENT RANDOM VARIABLES

6.2.1 Definitions and Construction Methods

K-wise independent random variables have been recently investigated extensively in connection with randomized algorithms in the field of theoretical computer science as well as with VLSI testing.

DEFINITION **6.12** Let q be a positive integer. A set of n random variables $x_1, x_2, ..., x_n$ in $I_q = \{0, 1, ..., q-1\}$ is said to be k-wise independent if for any $k(\leq n)$ positions $1 \leq j_1 < j_2 < \cdots < j_k \leq n$, we have

$$Pr[x_{j_1} = a_1, x_{j_2} = a_2, ..., x_{j_k} = a_k] = \frac{1}{q^k},$$

for all integers $(a_1, ..., a_k)$ with $0 \leq a_1, ..., a_k \leq q-1$. □

As was mentioned in Section 4.1, researchers in computational physics [21] working on high-precision Monte Carlo calculations, such as Ising model simulations, have a strong interest in pseudorandom number sequences that are immune to higher-order autocorrelation. It is easy to see that if a sequence of variables $x_1, x_2, ..., x_n$ in I_q is k-wise independent, the k-th order autocorrelation of the normalized sequence $u_i = x_i/q, i = 1, 2, ..., n$, becomes provably negligible.

Historically, the notion of pairwise independent but not mutually independent random variables dates back to Bernstein, followed by Lancaster, who presented methods of constructing such variables based on the Hadamard matrix and Latin squares [52]. Joffe [38] was the first person to give a method of constructing k-wise independent random variables in I_q for $q \geq n$. This is described in the following way:

Theorem 6.15 *For $n > 1$, let q be a prime power with $n \leq q(< 2n)$, and let $a_1, a_2, ..., a_k$ $(k \leq n)$ be random variables independently and uniformly chosen from $GF(q)$. Then*

$$x_i = \lambda(a_1 + a_2\psi(i) + a_2\psi(i)^2 + \cdots + a_k\psi(i)^{k-1}), \quad i = 1, ..., n,$$

are k-wise independent random variables in I_q, where the mappings, $\psi()$ and $\lambda()$, are bijections with $\psi(0) = \lambda(0) = 0$ between I_q and $GF(q)$, as defined in Section 6.1.

Proof: Consider the following Vandermonde matrix over $GF(q)$:

$$A = \begin{pmatrix} 1 & \psi(j_1) & \psi(j_1)^2 & \cdots & \psi(j_1)^{k-1} \\ 1 & \psi(j_2) & \psi(j_2)^2 & \cdots & \psi(j_2)^{k-1} \\ & \cdots & \cdots & \cdots & \\ & \cdots & \cdots & \cdots & \\ 1 & \psi(j_k) & \psi(j_k)^2 & \cdots & \psi(j_k)^{k-1} \end{pmatrix}.$$

Since $1 \leq j_1 < j_2 < \cdots < j_k \leq n \leq q$, all $\psi(j_i), i = 1, ..., k$, are distinct. Hence the matrix A is nonsingular. From the definitions of $x_1, ..., x_n$ and $a_1, ..., a_k$, the random variables $x_{j_1}, x_{j_2}, ..., x_{j_k}$ are independent and uniformly distributed in I_q for any k-tuple $(j_1, j_2, ..., j_k)$. □

The key advantage of the concept of k-wise independence lies in the fact that, as observed in the above construction, such variables rely on k random variables, $a_1, ..., a_k$, and thus the size of the probability space is q^k, while in the case of n mutually independent variables we need a probability space of a much larger size, namely, q^n, which is exponential in n. The next question is how much we can reduce the size of the probability space for k-wise independent random variables of length n. The following theorem gives the lower bound for the size of the space [3].

Theorem 6.16 *Any probability space with k-wise independent random variables in I_q of length n has a size $\Omega(n^{[k/2]})$.*

For $q = 2$, the best possible construction in the above sense can be obtained by using the binary BCH code in the field of coding theory [3]. Let $n = 2^m - 1$. Let $\beta_1, ..., \beta_n$ be n nonzero column vectors in $GF(2^m)$. Then consider the $(mt + 1) \times n$ matrix over $GF(2)$:

$$H = \begin{pmatrix} 1 & 1 & 1 & \cdots & 1 \\ \beta_1 & \beta_2 & \beta_3 & \cdots & \beta_n \\ \beta_1^3 & \beta_2^3 & \beta_3^3 & \cdots & \beta_n^3 \\ & \cdots & \cdots & \cdots & \\ \beta_1^{2t-1} & \beta_2^{2t-1} & \beta_3^{2t-1} & \cdots & \beta_n^{2t-1} \end{pmatrix},$$

which is the parity check matrix of the (primitive, narrow-sense) binary BCH code of designed distance $2t + 2 (= k + 1)$. Denote $\alpha = (a_1, ..., a_{mt+1})$ as a binary vector, and let $x_i = \langle \alpha, \mathbf{h}_i \rangle$ for $i = 1, ..., n$, where $\mathbf{h}_i$ is the ith column vector of the matrix H. If α is a randomly chosen vector, then the variables

$x_1, ..., x_n$ become k-wise independent random variables in $GF(2)$, because any k columns of the matrix H are linearly independent.

Denote the k-th order moment of a random variable y by $\tau_y^{(k)} = E[(y - \mu_y)^k]$, where $\mu_y = E(y)$. The next result ensures the importance of k-wise independence.

Proposition 6.2 *Let $x_1, ..., x_n$ be k-wise independent random variables. Let f be a real-valued function defined on the integers. Denote $y = \sum_{i=1}^n y_i$, where $y_i = f(x_i)$. Then we have*

$$\tau_y^{(k)} = \sum_{i=1}^{n} \tau_{y_i}^{(k)}.$$

Thus, for $k = 1$, i.e., no independence, we have $\mu_y = \sum_{i=1}^n \mu_{y_i}$, and for $k = 2$, i.e., pairwise independence, we have $\sigma_y^2 = \sum_{i=1}^n \sigma_{y_i}^2$. In what follows, the moment inequality also holds for k-wise independent random variables:

Proposition 6.3 *Let $x_1, ..., x_n$ be k-wise independent random variables. Let f be a real-valued function defined on the integers. Denote $y = \sum_{i=1}^n y_i$, where $y_i = f(x_i)$. If $\tau_y^{(k)}$ exists, then for an even k we have*

$$Pr[(y - \mu_y)^k > m] \leq \frac{\tau_y^{(k)}}{m}.$$

6.2.2 Almost K-wise Independence

As we showed in the previous section, for $q = 2$, the size of a k-wise independent probability space is bounded from below by approximately $n^{k/2}$, which means that a probability space of polynomial size in n can be constructed only for a fixed k. Thus, if $k = O(\log n)$, the size becomes sub-exponential. Here, by introducing the notion of *almost* k-wise independence, we further elaborate on the reduction of the space-size so that we obtain the probability space whose size is polynomial in n for every $k \leq n$.

DEFINITION **6.13** Let S_n be a sample space of n-bit strings, and let $x_1 \cdots x_n$ be an n-bit string uniformly chosen from S_n.

- S_n is said to be ε-away w.r.t. the L_∞-norm from k-wise independence, if for any k positions $1 \leq j_1 < j_2 < \cdots < j_k \leq n$ and any k-bit string s, we have

$$|Pr[x_{j_1}x_{j_2}\cdots x_{j_k} = \text{s}] - 2^{-k}| \leq \varepsilon.$$

- S_n is said to be ε-away w.r.t. the L_1-norm from k-wise independence, if for any k positions $1 \leq j_1 < j_2 < \cdots < j_k \leq n$ and any k-bit string s, we have

$$\sum_{\text{s}\in\{0,1\}^k} |Pr[x_{j_1}x_{j_2}\cdots x_{j_k} = \text{s}] - 2^{-k}| \leq \varepsilon.$$

□

Roughly speaking, every k-bit location in the almost k-wise independent sample string has a probability distribution that is statistically close to uniform. There are many possible choices of the metric for measuring the difference between the distribution induced by the sample space and the uniform distribution. The above definitions are the most typical. It is easy to see that if S_n is ε-away w.r.t. the L_∞-norm from k-wise independence, then it is at most $2^k\varepsilon$-away w.r.t. the L_1-norm. It is also easy to see that the notion of ε-away w.r.t. the L_∞-norm from k-wise independence is closely related to the notion of L_∞-discrepancy.

For any even k, the moment inequality is given as follows [66]:

Theorem 6.17 *Let S_n be a sample space of n-bit strings that is ε-away w.r.t the L_∞-norm from k-wise independence, and let $x_1 \cdots x_n$ be a string uniformly chosen from S_n. Let $y_i = (-1)^{x_i}, i = 1, ..., n$, and let $y = \sum_{i=1}^{n} y_i$. Note that $\mu_y = 0$. Then, for an even k, we have*

$$Pr[y^k > m] \leq \frac{B_k + \varepsilon n^k}{m},$$

where B_k is defined as the kth moment of the absolute value of the sum of n independent random variables that take their values from $\{-1, 1\}$ uniformly.

This result implies that our relaxation using the notion of almost k-wise independence does not have much effect on the properties relevant to the k-th order moments.

Before describing how to construct almost k-wise independent random variables, we first need to introduce the notion of linear tests.

DEFINITION **6.14** Let S_n be a sample space of n-bit strings, and let X be a string uniformly chosen from S_n. S_n is said to be ε-biased w.r.t. linear tests of size at most k if for every $\mathrm{s} \in \{0,1\}^n - \{0\}^n$ of weight at most k, we have

$$|Pr[\langle \mathrm{s}, X\rangle = 0] - Pr[\langle \mathrm{s}, X\rangle = 1]| \leq \varepsilon.$$

□

The following lemma, the so-called XOR lemma, is obtained from the finite, discrete version of Weyl's criterion for sequences of polynomials over $GF(2)$ (see Theorem 2.53).

Lemma 6.4 *If S_n is a sample space that is ε-biased w.r.t. linear tests of size k, then S_n is $(1-2^{-k})\varepsilon$-away w.r.t. the L_∞-norm, and $(2^k-1)^{1/2}\varepsilon$-away w.r.t. the L_1-norm from k-wise independence.*

We now show how to construct a sample space S_n that is ε-biased w.r.t. linear tests. N. Alon [4] presented two practical construction methods based on polynomial arithmetic over $GF(2)$. The first is

$$S_n^{(1)} = \{\langle \mathrm{vec}(\alpha^i), \mathrm{vec}(\beta)\rangle, i = 1, 2, ..., n\},$$

where α and β are uniformly chosen random elements in $GF(2^m)$, and vec() means the binary vector representation of elements in $GF(2^m)$. The next theorem evaluates the bias of this sample space $S_n^{(1)}$.

Theorem 6.18 *The sample space $S_n^{(1)}$ is ε-biased w.r.t. linear tests of size n, where $\varepsilon = \frac{n-1}{2^m}$.*

The second is

$$S_n^{(2)} = \{\ m\text{th-order LFSR sequences over } GF(2) \text{ of length } n\},$$

where the characteristic polynomial of LFSR sequences is a uniformly chosen irreducible polynomial of degree m and the initial values for the LFSR are uniformly chosen m bits. The next theorem evaluates the bias of this sample space $S_n^{(2)}$.

Theorem 6.19 *The sample space $S_n^{(2)}$ is ε-biased w.r.t. linear tests of size n, where $\varepsilon = \frac{n-1}{2^m}(1 + O(2^{-m/2}))$.*

From the above theorems, we get a sample space that is $(1-2^{-k})\varepsilon$-away w.r.t. the L_∞-norm, and $(2^k-1)^{1/2}\varepsilon$-away w.r.t. the L_1-norm from k-wise independence, for every $k \le n$. In this construction, we have $m = O(\log n + \log \varepsilon^{-1})$, that is, the size of the sample space becomes polynomial in n for every $k \le n$.

By using the parity check matrix of BCH codes described previously, J. Naor and M. Naor [66] found a general efficient scheme, which can be applied to ε-biased random variables to further reduce the size of the sample space. Let $\mathbf{h}_1, ..., \mathbf{h}_N$ be the column vectors of the parity check $(n \times N)$ matrix of BCH codes of designed distance $k+1$, where $n = 1 + \frac{k-1}{2}\log_2(N+1)$. The above authors obtained the following theorem:

Theorem 6.20 *If* $\mathbf{s}$ *is a sample from the* ε*-biased sample space* S_n *w.r.t. linear test of size* n, *then* $y_i = \langle \mathbf{s}, \mathbf{h}_i \rangle$, $i = 1, 2, ..., N$, *are* ε*-biased with respect to linear tests of size at most* k.

Proof: Let $V \subset \{1, 2, ..., N\}$ with $|V| = k$. For $b = 0, 1$, we have

$$Pr\left[\sum_{i\in V} y_i = b\right] = Pr\left[\left\langle \mathbf{s}, \sum_{i\in V} \mathbf{h}_i \right\rangle = b\right],$$

where the summation is done over $GF(2)$. Since any k column vectors of the parity check matrix are linearly independent, $\sum_{i\in V} \mathbf{h}_i$ is a nonzero n-bit pattern. From the assumption on the sample space S_n, we have

$$\left| Pr\left[\left\langle \mathbf{s}, \sum_{i\in V} \mathbf{h}_i \right\rangle = 0\right] - Pr\left[\left\langle \mathbf{s}, \sum_{i\in V} \mathbf{h}_i \right\rangle = 1\right] \right| \le \varepsilon,$$

which immediately leads us to

$$\left| Pr\left[\sum_{i\in V} y_i = 0\right] - Pr\left[\sum_{i\in V} y_i = 1\right] \right| \le \varepsilon.$$

□

Now, we summarize their method of constructing a sample space R_N of size $m = O(\log\log N + \log k + \log \varepsilon^{-1})$ bits, which is $(1-2^{-k})\varepsilon$-away w.r.t. the L_∞-norm, as well as $(2^k-1)^{1/2}\varepsilon$-away w.r.t. the L_1-norm from k-wise independence.

Algorithm for almost k-wise independent random variables
Step 1: Construct S_n of cardinality 2^m by using either of Alon's two algorithms
Step 2: Construct the sample space $R_N = \{y_1 \cdots y_N \mid y_i = \langle \mathrm{s}, \mathbf{h}_i \rangle, i = 1, 2, ..., N, \text{ for each } \mathrm{s} \in S_n\}$

* * *

Notes and Remarks

General discussions and detailed proofs on the discrepancy of (t, k)-sequences as well as on several applications of low-discrepancy sequences can be found in the excellent book by Niederreiter [72]. Generalized Niederreiter sequences were first introduced by Tezuka [91] in the analysis of polynomial arithmetic analogue of Halton sequences. Numerical integration problems in practical applications, such as financial engineering and computational physics, often have modest or large dimensions. Thus, the dependency of computing time on the dimension is a central issue for practitioners, although most theoretical results on discrepancy have so far provided only asymptotic bounds in terms of N, the number of points, assuming that the dimension k is fixed. The development of discrepancy theory explicitly stated in terms of both N and k is a very important challenge. For more information on this topic, see Wasilkowski and Woźniakowski [107].

By applying the notion of k-wise independence, we can accomplish full or partial derandomization of randomized algorithms for several problems in theoretical computer science, such as the maximal independent set problem, the set discrepancy problem, as well as Voronoi diagram construction and other problems in computational geometry [65]. Niederreiter and Schnorr [73] presented a method of constructing almost k-wise independent random variables for cryptographic applications.

A

SAMPLE C ROUTINES

The last task is to actually implement the algorithms.
– Henri Cohen (1993)

In this appendix, we describe three sample programs written in the ANSI C programming language: two subroutines for uniform random numbers, CombLS2 and CombMRG, which were implemented by using the Algorithm T in Section 4.1, for uniform random numbers, and another for low-discrepancy sequences, GFaure, which was implemented according to the algorithm based on Theorem 6.6.

Figure A.1 includes two subroutines: CombLS2Init and CombLS2, whose theoretical properties are given in detail in Example 4.7. CombLS2Init is an initialization routine. It requires as initial seeds three unsigned 32-bit integers, s1, s2, and s3, which should be positive integers less than 2^{28}, 2^{29}, and 2^{31}, respectively. We must call this routine once at the beginning of the main program before calling the main routine CombLS2, which produces a random number of 32-bit precision between 0 and 1 every time it is called in the program.

Figure A.2 shows two subroutines: CombMRGInit and CombMRG, which were theoretically investigated in Examples 4.8 and 4.13. Before calling an initialization routine, CombMRGInit, we have to provide two arrays of initial seeds, s1[0],...,s1[4] and s2[0],...,s2[6] with not all-zero unsigned 32-bit integers less than 2^{31} and 2^{29}, respectively. Like CombLS2Init, this routine must be called once at the beginning of the main program before the main routine CombMRG. Each call to CombMRG produces a random number of 32-bit precision between 0 and 1.

```
unsigned int s1, s2, s3;

void CombLS2Init() /* Extend s1, s2, s3 to 32-bit integers */
      {
       unsigned int b;

       b  = ((s1 << 9) ^ s1) << 4;
       s1 = (s1 << 4) ^ (b >> 28);
       b  = ((s2 << 2) ^ s2) << 3;
       s2 = (s2 << 3) ^ (b >> 29);
       b  = ((s3 << 6) ^ s3) << 1;
       s3 = (s3 << 1) ^ (b >> 31);
      }


double CombLS2() /* Generate numbers between 0 and 1. */
        {
         unsigned int b;

         b  = ((s1 << 9) ^ s1) << 4;
         s1 = (s1 << 13) ^ (b >> 19);
         b  = ((s2 << 2) ^ s2) << 3;
         s2 = (s2 << 20) ^ (b >> 12);
         b  = ((s3 << 6) ^ s3) << 1;
         s3 = (s3 << 17) ^ (b >> 15);
         return ((s1 ^ s2  ^ s3 ) * 2.3283064365e-10);
        }
```

Figure A.1 An implementation of CombLS2Init and CombLS2 in C.

Figures A.3 presents a subroutine for producing a generalized Faure sequence. This subroutine, GFaure, is defined and fully discussed in Section 6.1.2. Each call to GFaure outputs the next k-dimensional point in the sequence. This program requires two parameters, DIM and PR, which denote the dimension k and the smallest prime number no less than k, respectively, and the dimension k is assumed to be less than 500 in this program. Both parameters should be defined at the top of the main program. The input file 'GenMat.GF10' should contain the values of generator matrices for dimension 10 in this sample.

```
unsigned int s1[5], s2[7];

void CombMRGInit() /* Extend s1, s2, s3 to 32-bit integers */
 {
  unsigned int b;
  int k;

  for (k=0;k<5;k++){
   b = ((s1[k] << 3) ^ s1[k]) << 1;
   s1[k] = (s1[k] << 1) ^ (b >> 31);
  }

  for (k=0;k<7;k++){
   b = ((s2[k] << 2) ^ s2[k]) << 3;
   s2[k] = (s2[k] << 3) ^ (b >> 29);
  }
 }

double CombMRG()  /* Generate numbers between 0 and 1. */
 {
  static int j1=1, k1=4, j2=4, k2=6;
  unsigned int b1, b2;

  b1 = ((s1[k1] << 3) ^ s1[k1]) << 1;
  b2 = ((s1[j1] << 3) ^ s1[j1]) << 1;
  s1[k1] = (s1[k1]<<23) ^ (b1>>9) ^ (s1[j1]<<5) ^ (b2>>27);

  b1 = ((s2[k2] << 2) ^ s2[k2]) << 3;
  b2 = ((s2[j2] << 2) ^ s2[j2]) << 3;
  s2[k2] = (s2[k2]<<19) ^ (b1>>13) ^ (s2[j2]<<16) ^ (b2>>16);

  j1 -= 1; k1 -= 1; if(j1 < 0) j1 = 4; if(k1 < 0) k1 = 4;
  j2 -= 1; k2 -= 1; if(j2 < 0) j2 = 6; if(k2 < 0) k2 = 6;

  return ((s1[k1] ^ s2[k2]) * 2.3283064365e-10);
 }
```

Figure A.2 An implementation of CombMRGInit and CombMRG in C.

```
#define DIM 10
#define PR  11

GFaure(double x[500],int ix[500][35])
 {
  static int init=-1,mm,mp[35][500][35];
  int i,j,k,jj,save,pj[35];
  double xx;
  FILE *fp;

  if(init<0){
   mm=log(10000)/log(PR)+1;
   fp=fopen("GenMat.GF10","r");
   for(jj=1; jj<=DIM; jj++){
    for(i=0; i<mm; i++){
     for(j=0; j<mm ; j++){
      fscanf(fp,"%d",&mp[j][jj][i]);
     }
    }
   }
   init=0;
  }
  else{
   for(j=1; j<=DIM; j++){
    save=init; k=0;
    while((save%PR)==(PR-1)){k=k+1; save=save/PR;}
    xx=0;
    for(jj=mm-1; jj>=0; jj--){
     ix[j][jj]=(ix[j][jj]+mp[k][j][jj])%PR;
     xx=xx/PR+ix[j][jj];
    }
    x[j]=xx/PR;
   }
   init +=1;
  }
 }
```

Figure A.3 An implementation of GFaure in C.

REFERENCES

[1] L. Afflerbach and H. Grothe. The lattice structure of pseudorandom vectors generated by matrix generators. *Journal of Computational and Applied Mathematics*, 23:127–131, 1988.

[2] L. Afflerbach and K. Wenzel. Normal random numbers lying on spirals and clubs. *Statistische Hefte*, 29:237–244, 1988.

[3] N. Alon, L. Babai, and A. Itai. A fast and simple randomized parallel algorithm for the maximal independent set problem. *Journal of Algorithms*, 7:567–583, 1986.

[4] N. Alon, O. Goldreich, J. Håstad, and R. Peralta. Simple constructions of almost k-wise independent random variables. *Random Structures and Algorithms*, 3:289–304, 1992.

[5] D. A. André, G. L. Mullen, and H. Niederreiter. Figures of merit for digital multistep pseudorandom numbers. *Mathematics of Computation*, 54:737–748, 1990.

[6] I. A. Antonov and V. M. Saleev. An economic method of computing lp_τ-sequences. *USSR Computational Mathematics and Mathematical Physics*, 19:252–256, 1979.

[7] A.C. Arvillias and A.E. Maritsas. Partitioning the period of m-sequences and application to pseudorandom number generation. *Journal of the ACM*, 25:675–686, 1978.

[8] E. Bach. Realistic analysis of some randomized algorithms. *Journal of Computer and System Sciences*, 42:30–53, 1991.

[9] L. E. Baum and M. M. Sweet. Badly approximable power series in characteristic 2. *Annals of Mathematics*, 105:573–580, 1977.

[10] J. Beck and W. W. L. Chen. *Irregularities of Distribution.* Cambridge University Press, Cambridge, 1987.

[11] T. Beth and Z.D. Dai. On the complexity of pseudorandom sequences – or: If you can describe a sequence it can't be random. In J.-J. Quisquater and J. Vandewalle, editors, *Eurocrypt'89*, Lecture Notes in Computer Science, pages 533–543. Springer, 1989.

[12] M. Bicknell. A primer for the Fibonacci numbers: Part VII. *The Fibonacci Quarterly*, 8:407–420, 1970.

[13] L. Blum, M. Blum, and M. Shub. A simple unpredictable pseudorandom number generator. *SIAM Journal on Computing*, 15:364–383, 1986.

[14] M. Blum and S. Micali. How to generate cryptographically strong sequences of pseudorandom bits. *SIAM Journal on Computing*, 13:850–864, 1984.

[15] E. Braaten and G. Weller. An improved low-discrepancy sequence for multidimensional quasi-Monte Carlo integration. *Journal of Computational Physics*, 33:249–258, 1979.

[16] P. Bratley, B. L. Fox, and L. E. Schrage. *A Guide to Simulation*. Springer-Verlag, New York, 2nd edition, 1987.

[17] R. P. Brent. On the periods of generalized Fibonacci recurrences. *Mathematics of Computation*, 63:389–401, 1994.

[18] G. J. Chaitin. *Information, Randomness & Incompleteness. Papers on Algorithmic Information Theory.* World Scientific Publishers, Singapore, 1987.

[19] P. D. Coddington. Analysis of random number generators using Monte Carlo simulation. *International Journal of Modern Physics*, 5:547–560, 1994.

[20] H. Cohen. *A Course in Computational Algebraic Number Theory.* Graduate Texts in Mathematics. Springer-Verlag, Berlin, 1993.

[21] A. Compagner. The hierarchy of correlations in random binary sequences. *Journal of Statistical Physics*, 63:883–896, 1991.

[22] R. Couture, P. L'Ecuyer, and S. Tezuka. On the distribution of k-dimensional vectors for simple and combined Tausworthe sequences. *Mathematics of Computation*, 60:749–761, 1993.

[23] R. R. Coveyou and R. D. MacPherson. Fourier analysis of uniform random number generators. *Journal of the ACM*, 14:100–119, 1967.

[24] L. Devroye. *Non-Uniform Random Variate Generation.* Springer-Verlag, New York, 1986.

[25] J. Eichenauer and J. Lehn. A nonlinear congruential pseudorandom number generators. *Statistische Hefte*, 27:315–326, 1986.

[26] J. Eichenauer and H. Niederreiter. On Marsaglia's lattice test for pseudorandom numbers. *Manuscripta Mathematica*, 62:245–248, 1988.

[27] J. Eichenauer-Herrmann. Inversive congruential pseudorandom numbers avoid the planes. *Mathematics of Computation*, 56:297–301, 1991.

[28] A.M. Ferrenberg, D.P. Landau, and Y.J. Wong. Monte Carlo simulations: Hidden errors from "good" random number generators. *Physical Review Letters*, 69:3382–3384, 1992.

[29] U. Fincke and M. Pohst. Improved methods for calculating vectors of short length in a lattice. *Mathematics of Computation*, 44:463–471, 1985.

[30] M. Fushimi and S. Tezuka. The k-distribution of generalized feedback shift register pseudorandom numbers. *Communications of the ACM*, 26:516–523, 1983.

[31] P.M. Gruber and C.G. Lekkerkerker. *Geometry of Numbers.* North-Holland, Amsterdam, 2nd edition, 1987.

[32] J.H. Halton. On the efficiency of certain quasirandom sequences of points in evaluating multidimensional integrals. *Numerische Mathematik*, 2:84–90, 1960.

[33] G. H. Hardy and E. M. Wright. *An Introduction to the Theory of Numbers.* Oxford University Press, Oxford, 5th edition, 1983.

[34] J.R. Heringa, H.W.J. Blöte, and A. Compagner. New primitive trinomials of Mersenne-exponent degrees for random-number generation. *International Journal of Modern Physics C*, 3:561–564, 1992.

[35] Y.C. Ho and X.R. Cao. *Perturbation Analysis of Discrete Event Dynamic Systems.* Kluwer Academic Publishers, 1991.

[36] A. Hoogland, J. Spaa, B. Selman, and A. Compagner. A special-purpose processor for the Monte Carlo simulation of Ising spin systems. *Journal of Computational Physics*, 51:250–260, 1983.

[37] R. Impagliazzo, L. Levin, and M. Luby. Pseudorandom generation from one-way functions. In *Proceedings of the 21st Annual ACM Symposium on Theory of Computing*, pages 12–24. ACM, 1989.

[38] A. Joffe. On a set of almost deterministic k-independent random variables. *Annals of Probability*, 2:161–162, 1974.

[39] M.H. Kalos and P.A. Whitlock. *Monte Carlo Methods*, volume I. John Wiley & Sons, New York, 1986.

[40] D. E. Knuth. *The Art of Computer Programming: Seminumerical Algorithms*, volume 2. Addison-Wesley, 2nd edition, 1981.

[41] A. N. Kolmogorov. Three approaches to defining "quantity of information". *Problems Information Transmission*, 1:3–11, 1965.

[42] R.F. Koopman. The orders of equidistribution of subsequences of some asymptotically random sequences. *Communications of the ACM*, 29:802–806, 1986.

[43] H. Krawczyk. How to predict congruential generators. *Journal of Algorithms*, 13:527–545, 1992.

[44] L. Kuipers and H. Niederreiter. *Uniform Distribution of Sequences.* John Wiley & Sons, New York, 1974.

[45] P. L'Ecuyer, F. Blouin, and R. Couture. A search for good multiple recursive random number generators. *ACM Transactions on Modeling and Computer Simulation*, 3:87–98, 1993.

[46] P. L'Ecuyer and S. Tezuka. Structural properties for two classes of combined random number generators. *Mathematics of Computation*, 57:735–746, 1991.

[47] A. K. Lenstra. Factoring multivariate polynomials over finite fields. *Journal of Computer System and Science*, 30:235–248, 1985.

[48] A. K. Lenstra, Jr. H. W. Lenstra, and L. Lovász. Factoring polynomials with rational coefficients. *Mathematische Annalen*, 261:515–534, 1982.

[49] P. A. W. Lewis, A. S. Goodman, and J. M. Miller. A pseudorandom number generator for the System/360. *IBM System Journal*, 8:136–146, 1969.

[50] T. G. Lewis and W. H. Payne. Generalized feedback shift register pseudorandom number algorithms. *Journal of the ACM*, 20:456–468, 1973.

[51] R. Lidl and H. Niederreiter. *Introduction to Finite Fields and Their Applications.* Cambridge University Press, Cambridge, 1986.

[52] M. Luby. A simple parallel algorithm for the maximal independent set problem. *SIAM Journal on Computing*, 15:1036–1053, 1986.

[53] K. Mahler. An analogue to Minkowski's geometry of numbers in a field of series. *Annals of Mathematics*, 42:488–522, 1941.

[54] K. Mahler. On a theorem in the geometry of numbers in a space of Laurent series. *Journal of Number Theory*, 17:403–416, 1983.

[55] G. Marsaglia. Random numbers fall mainly in the planes. *Proceedings of National Academy of Sciences*, 61:25–28, 1968.

[56] G. Marsaglia. Random number generation. In A. Ralston and C.L. Meek, editors, *Encyclopedia of Computer Science*, pages 1192–1197. Petrocelli/Charter, New York, 1976.

[57] G. Marsaglia. A current view of random number generators. In L. Billard, editor, *Computer Science and Statistics: the 16-th Symposium on the Interface*, pages 3–10. Elsevier Science (North-Holland), 1985.

[58] G. Marsaglia. Remarks on choosing and implementing random number generators. *Communications of the ACM*, 36:105–108, 1993.

[59] G. Marsaglia and A. Zaman. A new class of random number generators. *Annals of Applied Probability*, 1:462–480, 1991.

[60] M. Matsumoto and Y. Kurita. Twisted GFSR generators. *ACM Transactions on Modeling and Computer Simulation*, 2:179–194, 1992.

[61] J. P. Mesirov and M. M. Sweet. Continued fraction expansions of rational expressions with irreducible denominators in characteristic 2. *Journal of Number Theory*, 27:144–148, 1987.

[62] W. J. Morokoff and R. E. Caflisch. Quasi-random sequences and their discrepancies. *SIAM Journal on Scientific Computing*, 15:1251–1279, 1994.

[63] R. Motwani and P. Raghavan. *Randomized Algorithms.* Cambridge University Press, Cambridge, 1995.

[64] G. L. Mullen and H. Niederreiter. Optimal characteristic polynomials for digital multistep pseudorandom numbers. *Computing*, 39:155–163, 1987.

[65] K. Mulmuley. Randomized geometric algorithms and pseudorandom generators. In *Proceedings of the 33th Symposium on Foundations of Computer Science*, pages 90–100. IEEE, 1992.

[66] J. Naor and M. Naor. Small-bias probability space: Efficient construction and applications. *SIAM Journal on Computing*, 22:838–856, 1993.

[67] H. R. Neave. On using the box-muller transformation with multiplicative congruential pseudorandom number generators. *Applied Statistics*, 22:92–97, 1973.

[68] H. Niederreiter. Quasi-Monte Carlo methods and pseudorandom numbers. *Bulletins of American Mathematical Society*, 84:957–1041, 1978.

[69] H. Niederreiter. Point sets and sequences with small discrepancy. *Monatshefte für Mathematik*, 104:273–337, 1987.

[70] H. Niederreiter. Low-discrepancy and low-dispersion sequences. *Journal of Number Theory*, 30:51–70, 1988.

[71] H. Niederreiter. The probabilistic theory of linear complexity. In C. G. Günther, editor, *Eurocrypt'88*, volume 330 of *Lecture Notes in Computer Science*, pages 191–209. Springer, 1988.

[72] H. Niederreiter. *Random Number Generation and Quasi-Monte Carlo Methods.* Number 63 in CBMS-NSF Regional Conference Series in Applied Mathematics. SIAM, 1992.

[73] H. Niederreiter and C. P. Schnorr. Local randomness in polynomial random number and random function generators. *SIAM Journal on Computing*, 22:684–694, 1993.

[74] R.B. Pearson, J.L. Richardson, and D. Toussaint. A fast processor for Monte-Carlo simulation. *Journal of Computational Physics*, 51:241–249, 1983.

[75] C.K. Peng, S. Prakash, H.J. Herrmann, and H.E. Stanley. Randomness versus deterministic chaos: Effect on invasion percolation clusters. *Physical Review A*, 42:4537–4542, 1990.

[76] M. Pohst and H. Zassenhaus. *Algorithmic Algebraic Number Theory.* Encyclopedia of Mathematics and its Applications. Cambridge University Press, Cambridge, 1989.

[77] B. D. Ripley. The lattice structure of pseudorandom number generators. *Proceedings of Royal Society of London, Series A*, 389:197–204, 1983.

[78] B. D. Ripley. *Stochastic Simulation.* John Wiley & Sons, New York, 1987.

[79] R. Rivest, A. Shamir, and L. Adleman. A method for obtaining digital signatures and public-key cryptosystems. *Communications of the ACM*, 21:120–126, 1978.

[80] K.F. Roth. On irregularities of distribution. *Mathematika*, 1:73–79, 1954.

[81] K.F. Roth. On irregularities of distribution. IV. *Acta Arithmetica*, 37:67–75, 1980.

[82] R. A. Rueppel. *Analysis and Design of Stream Ciphers*. Communications and Control Engineering. Springer-Verlag, Berlin, 1986.

[83] C.P. Schnorr. A more efficient algorithm for lattice basis reduction. *Journal of Algorithms*, 9:47–62, 1988.

[84] I.M. Sobol'. The distribution of points in a cube and the approximate evaluation of integrals. *USSR Computational Mathematics and Mathematical Physics*, 7:86–112, 1967.

[85] R. C. Tausworthe. Random numbers generated by linear recurrence modulo two. *Mathematics of Computation*, 19:201–209, 1965.

[86] S. Tezuka. On the discrepancy of GFSR pseudorandom numbers. *Journal of the ACM*, 34:939–949, 1987.

[87] S. Tezuka. Walsh-spectral test for GFSR pseudorandom number generators. *Communications of the ACM*, 30:731–735, 1987.

[88] S. Tezuka. On optimal GFSR pseudorandom number generators. *Mathematics of Computation*, 50:531–533, 1988.

[89] S. Tezuka. Lattice structure of pseudorandom sequences from shift register generators. In R. E. Nance, editor, *the 1990 Winter Simulation Conference*, pages 266–269. IEEE, 1990.

[90] S. Tezuka. Neave effect also occurs with Tausworthe sequences. In G. M. Clark, editor, *the 1991 Winter Simulation Conference*, pages 1030–1034. IEEE, 1991.

[91] S. Tezuka. Polynomial arithmetic analogue of Halton sequences. *ACM Transactions on Modeling and Computer Simulation*, 3:99–107, 1993.

[92] S. Tezuka. The k-dimensional distribution of combined GFSR sequences. *Mathematics of Computation*, 62:809–817, 1994.

[93] S. Tezuka. A unified view of large-period random number generators. *Journal of Operations Research Society of Japan*, 37:211–227, 1994.

[94] S. Tezuka and M. Fushimi. Fast generation of low-discrepancy points based on Fibonacci polynomials. In J. R. Wilson, editor, *the 1992 Winter Simulation Conference*, pages 433–437. IEEE, 1992.

[95] S. Tezuka and M. Fushimi. Calculation of Fibonacci polynomials for GFSR sequences with low discrepancies. *Mathematics of Computation*, 60:763–770, 1993.

[96] S. Tezuka and P. L'Ecuyer. Efficient and portable combined Tausworthe random number generators. *ACM Transactions on Modeling and Computer Simulation*, 1:99–112, 1991.

[97] S. Tezuka, P. L'Ecuyer, and R. Couture. On the lattice structure of the add-with-carry and subtract-with-borrow random number generators. *ACM Transactions on Modeling and Computer Simulation*, 3:315–331, 1993.

[98] S. Tezuka and T. Tokuyama. A note on polynomial arithmetic analogue of Halton sequences. *ACM Transactions on Modeling and Computer Simulation*, 4:279–284, 1994.

[99] S. Tezuka and Y. J. Wong. Ising model simulations with two classes of random number generators. Technical Report RT-0104, IBM Research, Tokyo Research Laboratory, 1994.

[100] J. P. R. Tootill, W. D. Robinson, and A. G. Adams. The runs up-and-down performance of Tausworthe pseudorandom number generators. *Journal of the ACM*, 18:381–399, 1971.

[101] J. P. R. Tootill, W. D. Robinson, and D. J. Eagle. An asymptotically random Tausworthe sequence. *Journal of the ACM*, 20:469–481, 1973.

[102] J.F. Traub and H. Woźniakowski. The Monte Carlo algorithm with a pseudorandom generator. *Mathematics of Computation*, 58:323–339, 1992.

[103] A. J. van de Goor. *Testing Semiconductor Memories: Theory and Practice.* John Wiley & Sons, Chichester, 1991.

[104] J. von Neumann. Various techniques used in connection with random digits. In *Collected Works*, volume 5, pages 768–770. Pergamon Press, Oxford, 1963.

[105] D. Wang and A. Compagner. On the use of reducible polynomials as random number generators. *Mathematics of Computation*, 60:363–374, 1993.

[106] T. T. Warnock. Computational investigations of low-discrepancy point sets. In S. K. Zaremba, editor, *Applications of Number Theory to Numerical Analysis*, pages 319–343. Academic Press, New York, 1972.

[107] G. W. Wasilkowski and H. Woźniakowski. Explicit cost bounds of algorithms for multivariate tensor product problems. *Journal of Complexity*, 11:to appear, March 1995.

[108] B. A. Wichmann and I. D. Hill. An efficient and portable pseudorandom number generator. *Applied Statistics*, 31:188–190, 1982. (See also corrections and remarks in the same journal by Wichmann and Hill 33 (1984), 123; McLeod 34 (1985), 198-200; Zeisel 35 (1986), 89.).

[109] H. Woźniakowski. Average case complexity of multivariate integration. *Bulletin of the American Mathematical Society*, 24:185–194, 1991.

[110] A. Yao. Theory and applications of trapdoor functions. In *Proceedings of the 23rd Annual Symposium on Foundations of Computer Science*, pages 80–91. IEEE, 1982.

[111] S. K. Zaremba. The mathematical basis of Monte Carlo and quasi-Monte Carlo methods. *SIAM Review*, 10:303–314, 1968.

INDEX